The *SUPERPOWER* Field Guide

OSTRICHES

BY RACHEL POLIQUIN

ILLUSTRATED BY
NICHOLAS JOHN FRITH

HOUGHTON MIFFLIN HARCOURT
Boston New York

For Tobias, of course, and always. –R.P.

To the amazing bird in my life, and our egg of wonder. –N.J.F.

A great big ostrich-size thank you to Dr. John Hutchinson, Professor of Evolutionary Biomechanics, Royal Veterinary College, University of London.

hmhbooks.com

The illustrations in this book were produced using a mixture of black ink, pencil, and wax crayon on paper, in a technique known as "preseparation." For printing purposes here, the artwork was colored digitally.

The text type was set in Adobe Caslon Pro.
The display type was set in Sign Painter House Showcard.

ISBN: 978-0-544-95040-5

Manufactured in China
SCP 10 9 8 7 6 5 4 3 2 1
4500769241

HAKUNA MATATA!
The SUPERPOWER Field Guide
OSTRICHES
Illustrated by NICHOLAS JOHN FRITH
RACHEL POLIQUIN
THE ORIGIN OF SPECIES
DARWIN
ATTENBOROUGH
MOLES
EELS
BEAVERS

THIS AN OSTRICH.

Just an ordinary ostrich.

But even ordinary ostriches are extraordinary. In fact, even ordinary ostriches are superheroes.

I know what you're thinking. You're thinking that ostriches are just overgrown chickens with ridiculous necks, skinny legs, and bad attitudes.

And you're right! Believe it or not, that neck helps ostriches run at supersonic speeds. Those skinny legs can kill a lion dead. And that bad attitude? Well now, you can't be one of the biggest, fastest, fiercest warriors around without having a touch of grumpy swagger.

You're still not convinced that ostriches are superpowered, are you?

Well, you don't know ostriches.

But you will.

MEET ENO

MEET ENO, AN ORDINARY OSTRICH. He's big. He's grumpy. He has the longest legs and the tiniest head, and you should never, ever underestimate what Eno can do. His superpowers include:

COLOSSAL ORBS OF TELESCOPIC VISION
(That means Eno has really big eyeballs.)

THIGHS OF THUNDER

TOE CLAWS OF DEATH

SUPER-FANTASTIC ELASTIC STRIDERS

TWO-TOED TORPEDOES

DO-IT-ALL DINO FLAPS

THE IMPOSSIBLE EVER-FLOW LUNG

EPIC ENDURANCE

THE EGG OF WONDER

HYDRO-HOARDING HEAT SHIELD

Are you ready? Are you brave enough? Because this battle bird is built ferociously fierce! Unfathomably fast! So take a deep breath. It's time to meet ***ENO, SUPERSONIC SURVIVOR!***

ENO HAS HORRIBLE FEET

THE FIRST THING YOU NEED TO KNOW about Eno is that he has horrible feet. They're scabby and rough with scaly plates down his toes. He only has two toes on each foot, and they are both horrible. His big toe has a giant black toenail that's longer than a grizzly bear's claw. His other toe is just a knob on the side of his foot. It doesn't have a toenail, and it's ugly.

I know it's not nice to say such things. But look at that foot—you know it's true! Besides, I think Eno is proud of his horrible feet. Those feet are survivor feet, and they save his life every day in one of the toughest, hottest, fiercest places on the planet.

You see, Eno lives in part of the African ***savanna,*** in Tanzania. Although it is a beautiful place, it can also be a terrible place. In the dry season, the hot sun bakes the ground until it cracks. Water can be hard to find, and Eno might go days, even weeks, without a drink. But water isn't Eno's only problem. The savanna is also home to lions, cheetahs, leopards, and hyenas—some of the hungriest, wildest, fastest predators on earth.

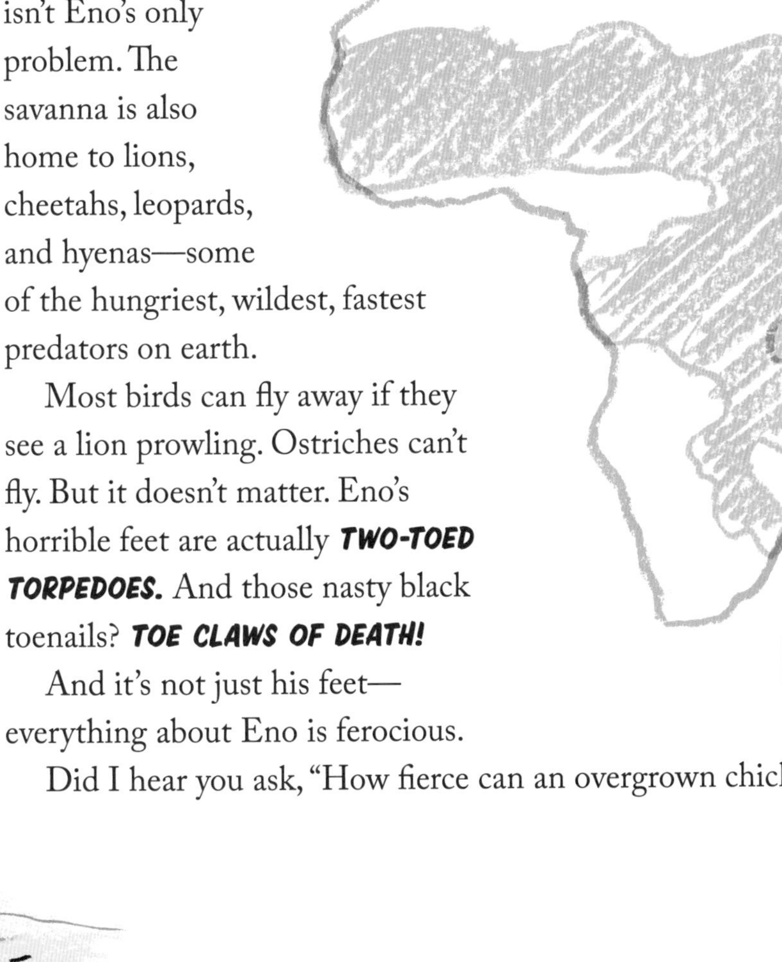

Most birds can fly away if they see a lion prowling. Ostriches can't fly. But it doesn't matter. Eno's horrible feet are actually ***TWO-TOED TORPEDOES.*** And those nasty black toenails? ***TOE CLAWS OF DEATH!***

And it's not just his feet—everything about Eno is ferocious.

Did I hear you ask, "How fierce can an overgrown chicken really be?"

Well, to understand just how fierce Eno is, you need to understand where Eno lives. And to understand where Eno lives, we need to go. You and me. Because the African savanna is more than just a place on a map. It's a way of life. And it sharpens and shapes every inch of Eno into a superfierce survivor.

I know we've just met, but we should leave. Right now. Not a minute to lose. The dry season has started. The wildebeests are already on the move, and life is getting gritty. Don't pack much. Just bring a T-shirt, shorts, a large hat, sunscreen, water, and binoculars. I'll bring the sandwiches.

THE WATCHING GAMES

WELCOME TO PARADISE! We're in the Serengeti, a small corner of the African savanna in Tanzania. You might be far from home, but I bet you've seen this place thousands of times in pictures: grasslands stretching as far as you can see, dotted here and there with graceful acacia trees.

It's home to some of the world's most splendid animals: elephants, giraffes, zebras, wildebeests, lions, hyenas, gazelles, cheetahs, impalas, rhinoceros, warthogs, and, of course, ostriches. They're all here, roaming the grassy plains of the African savanna.

If we had come in April, it would be raining. During the rainy season, it rains most days. With months of heavy rain, grasses grow tall and lush. Rivers flood and pool into huge watering holes. There is lots to eat and lots to drink—life is good in the wet season.

But we've come in late September—the peak of the dry season—and it hasn't really rained for months. The grasses are shriveled and burnt. The watering holes have shrunk to mean and muddy puddles. Everyone fights for a drink. Some animals, like the wildebeests, have wandered north to find food and water. Life is tough in the dry season.

But whether it's the dry season or the wet season, one thing is always the same in the Serengeti: ***EVERYONE IS ALWAYS WATCHING EVERYONE ELSE.***

Of course, wild animals are always on the lookout. Most prefer to sneak among the trees or lurk in the shadows. But on the wide-open grassy plains of the African savanna, there is nowhere to hide. Maybe slinky things like snakes and mongooses can creep by unnoticed, and even a cheetah might slip into the tall grass during the rainy season. But when dry season hits, everybody can always see everybody else.

WET SEASON
DRY SEASON

The lions are always watching the zebras, sometimes from atop a termite mound, so the zebras can always see the lions always watching them.

The gazelles are always watching the cheetahs, and the cheetahs are always watching the gazelles . . .

and the hyenas are watching the cheetahs as they watch the gazelles, in the hopes of getting a bit of dinner.

The ostriches are always watching the lions and cheetahs. Giraffes are so tall, they are very good at watching everyone. And don't think they aren't all watching us as we watch them.

Now, what everyone is really doing is waiting for someone to make a move. For a zebra to stray a little too far from the herd. For a cheetah to creep a few steps forward. And suddenly the lazy-looking savanna springs into high-speed action. But until then, everyone is just watching and waiting. And maybe munching some grass.

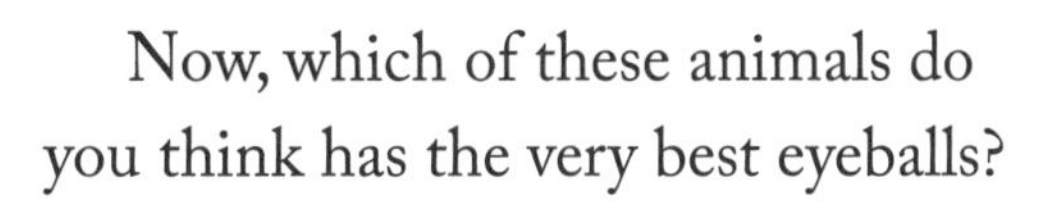

Now, which of these animals do you think has the very best eyeballs?

I hope you guessed Eno. Because Eno's eyeballs are the first weapon in his **ARSENAL OF SUPERFIERCE SURVIVAL SKILLS.** Do you know the word *arsenal?* It means a collection of weapons and tools. An army has tanks and guns in its arsenal. Pirates have cannons and cutlasses. One of the most important parts of any defense is the lookout—you always want to spot the enemy before they spot you. Generals have binoculars. Pirates have spyglasses. And Eno has **COLOSSAL ORBS OF TELESCOPIC VISION.**

SUPERPOWER #1

COLOSSAL ORBS OF TELESCOPIC VISION

HERE'S AN INTERESTING FACT: birds have far, far better vision than we do, or than any other mammal does. Birds can see colors we can't even imagine. They see movements that are too fast for our eyes to notice. Their left and right eyes can see different things at the same time! Even scientists don't yet understand all the amazingness of bird eyeballs.

Part of the reason birds see so much better is they have huge eyes. An eagle or an owl is less than a third of your size but has eyeballs as big as yours. And Eno? Well, Eno has the biggest eyeballs of any animal on land. Let me say that again: **ENO HAS THE BIGGEST EYEBALLS OF ANY ANIMAL ON LAND.** His eyeballs are bigger than a giraffe's or even an elephant's! Only sea giants like whales and colossal squids have bigger eyes.

Bigger eyes mean better vision. But before I explain why, do you understand how eyeballs work? Eyeballs are amazing and intricate and very complicated. But an eyeball is basically just a ball of jelly with a hole at the front and a screen at the back. Light goes in the hole and shows an image of the outside world on the screen. Then special light-gathering information bits inside the screen send all the information about that image to your brain.

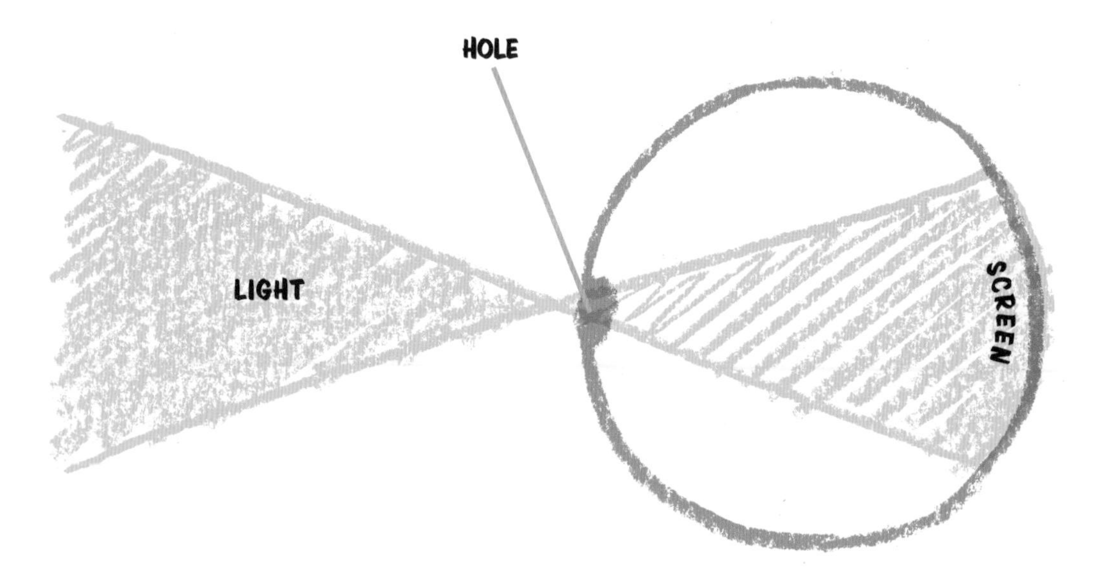

As you probably know with televisions, a bigger screen usually has a better image. A giant television screen has so many more pixels—or dots of color and detail—than the teeny TV in your granny's kitchen, which means a giant screen would show your granny's cooking shows in way more detail, if she ever bought a new TV.

The same goes for eyes. The huge screens at the back of Eno's colossal eyeballs have way, way more light-gathering information bits than the small screens in your eyes, which means Eno sees the world in much, much clearer detail than you do. Eno's eyes can spot the slightest twitch of grass that our eyeballs wouldn't notice. That might give him a split-second head start that could save his life. And more of those light-gathering information bits also means Eno sees much better in dimmer light at dusk when lions like to hunt.

Now, your teacher will be angry if I don't tell you the proper words for your eyeball parts. So, the hole at the front is called a ***pupil***. The screen at the back is a ***retina***. The jelly is called ***vitreous humor***—it looks like colorless hair gel. And the light-gathering information bits are called ***photoreceptors***. There are different kinds of photoreceptors that do different things, but we don't need to worry about that right now.

Bird eyes also have something yours don't—a strange comb-like lump filled with blood vessels. It's inside their eyeball, and it's called a ***pecten*** or a *pecten oculi*, if you want to be fancy. Scientists don't understand all the amazing things it does, but they know it nourishes the retina. And that means the retina needs fewer blood vessels, so there is room in the retina for even more photoreceptors, for even sharper bird vision. Wow!

Next, Eno's eyeballs are shaped differently than yours. Your eyeball is rounder at the back, which means the image quickly gets blurry as the retina curves around the sides of your eyeball. But ostrich eyes are flatter at the back, which means the retina is flatter, which means more of the image is in focus, especially at the sides. And that means Eno could read this book just as well if you held it at the side of his head as if you held it straight in front of him.

Birds not only have more photoreceptors than you, they have different kinds that let them see colors we can't even imagine. They also have colored oil drops in their photoreceptors, which might let them see subtle differences between colors. Birds can even see ultraviolet light, which is completely invisible to us. It's a rainbow world for birds!

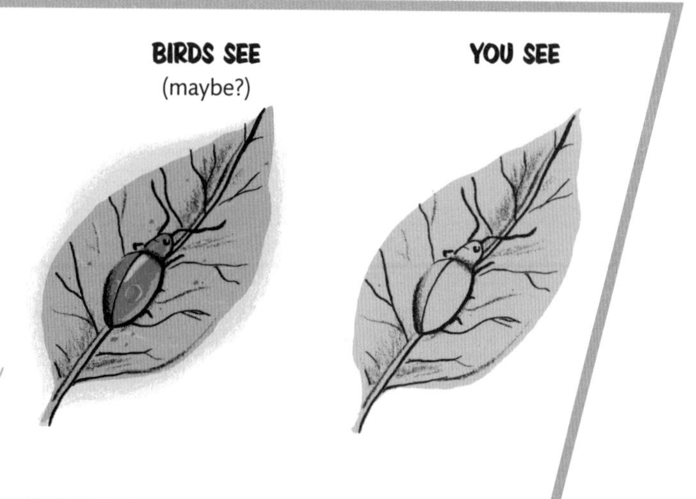

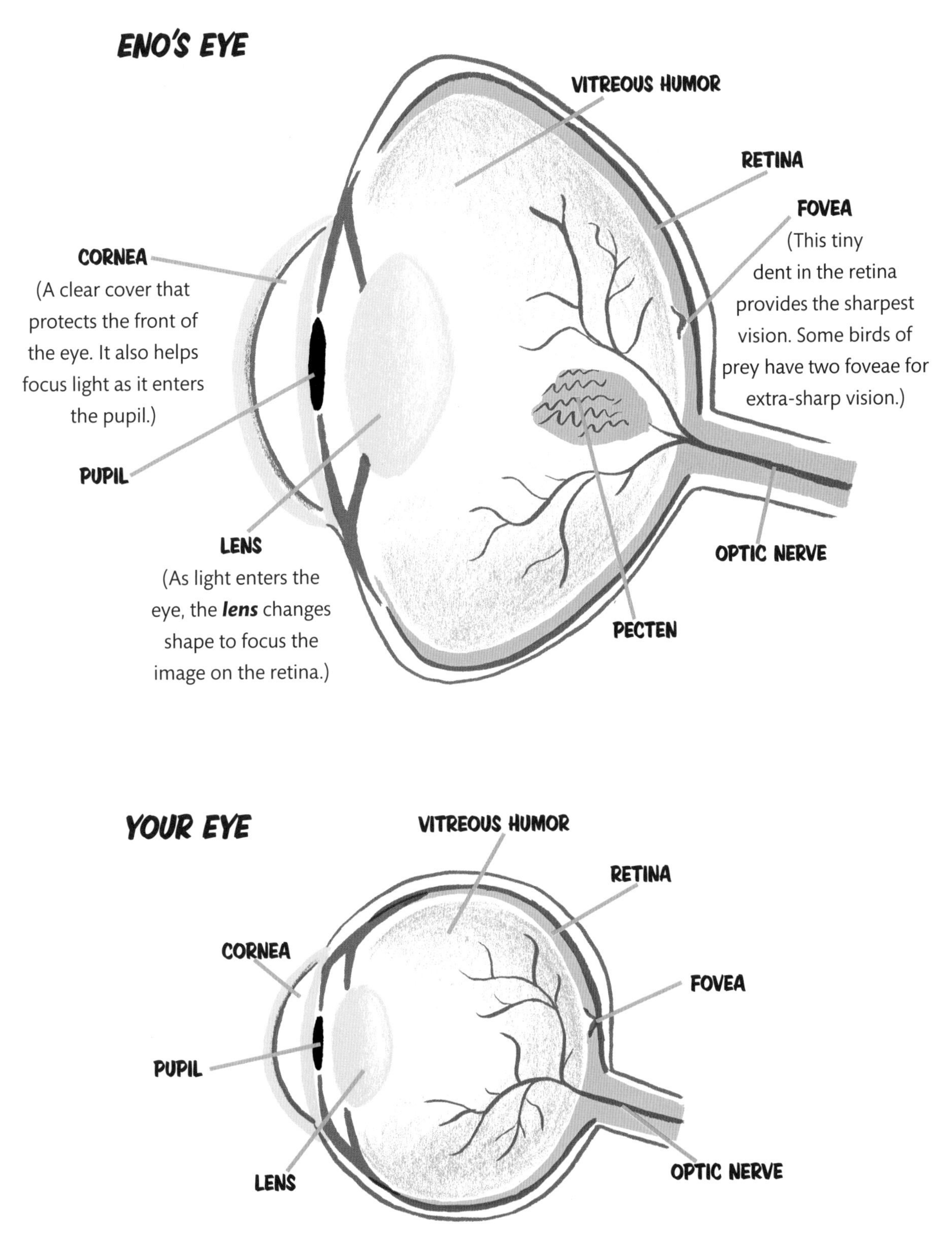

*EYES ARE RELATIVE IN SIZE

Hmm . . . ignore that last thing I said. It's a bad example, and not only because Eno can't read. Eno's eyeballs are on the sides of his head—not at the front like yours—so his sideways vision is excellent. (Scientists call sideways vision ***peripheral vision***.) In fact, because his eyes are on the sides, if you held this book right in front of Eno, he might not be able to see it at all. Ostriches can't see their own beaks!

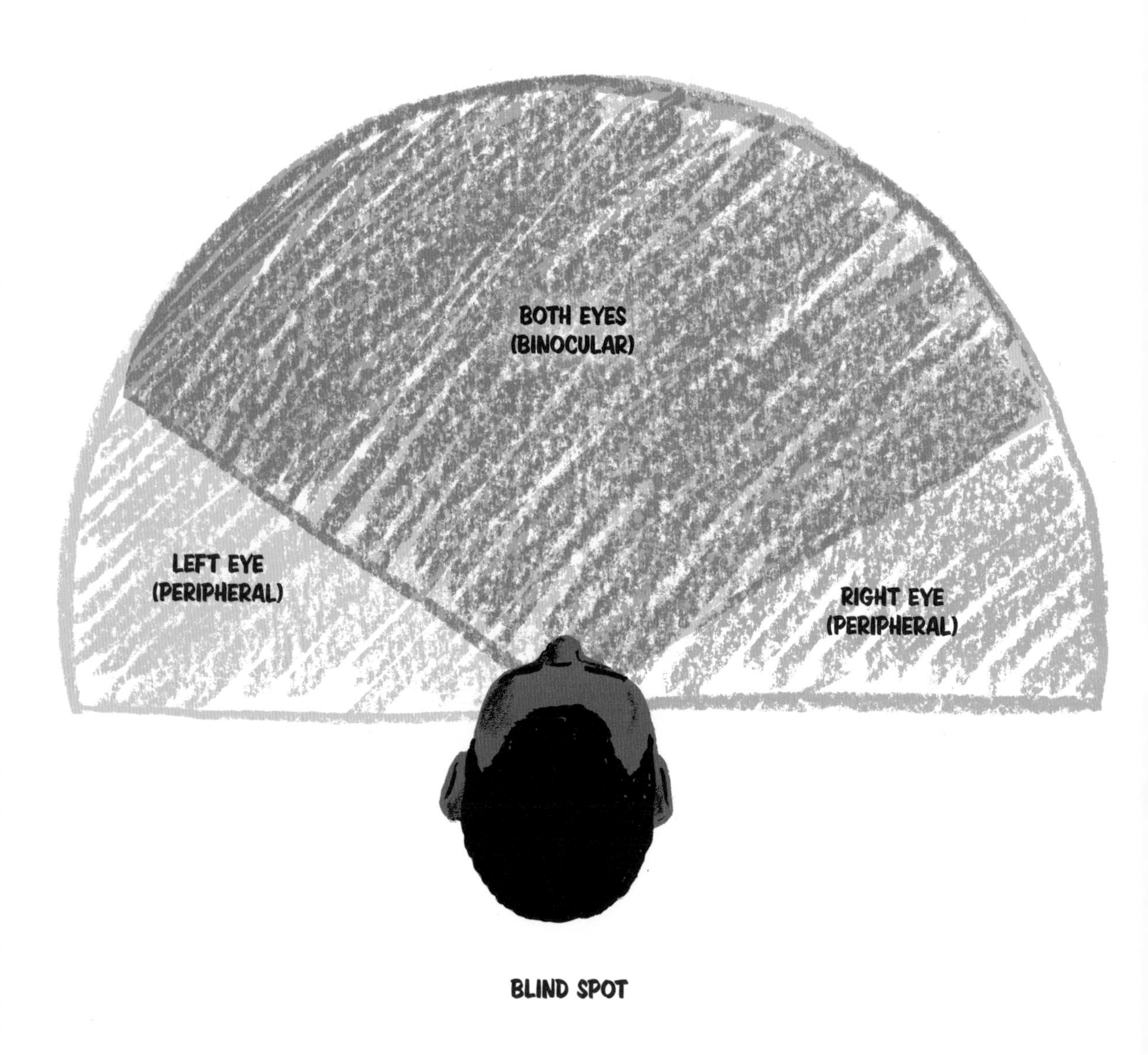

They can't even see what they are pecking at the moment they're pecking it! Crazy! But ostriches can see almost all the way around their heads, which is incredibly helpful when you're always on the lookout for predators.

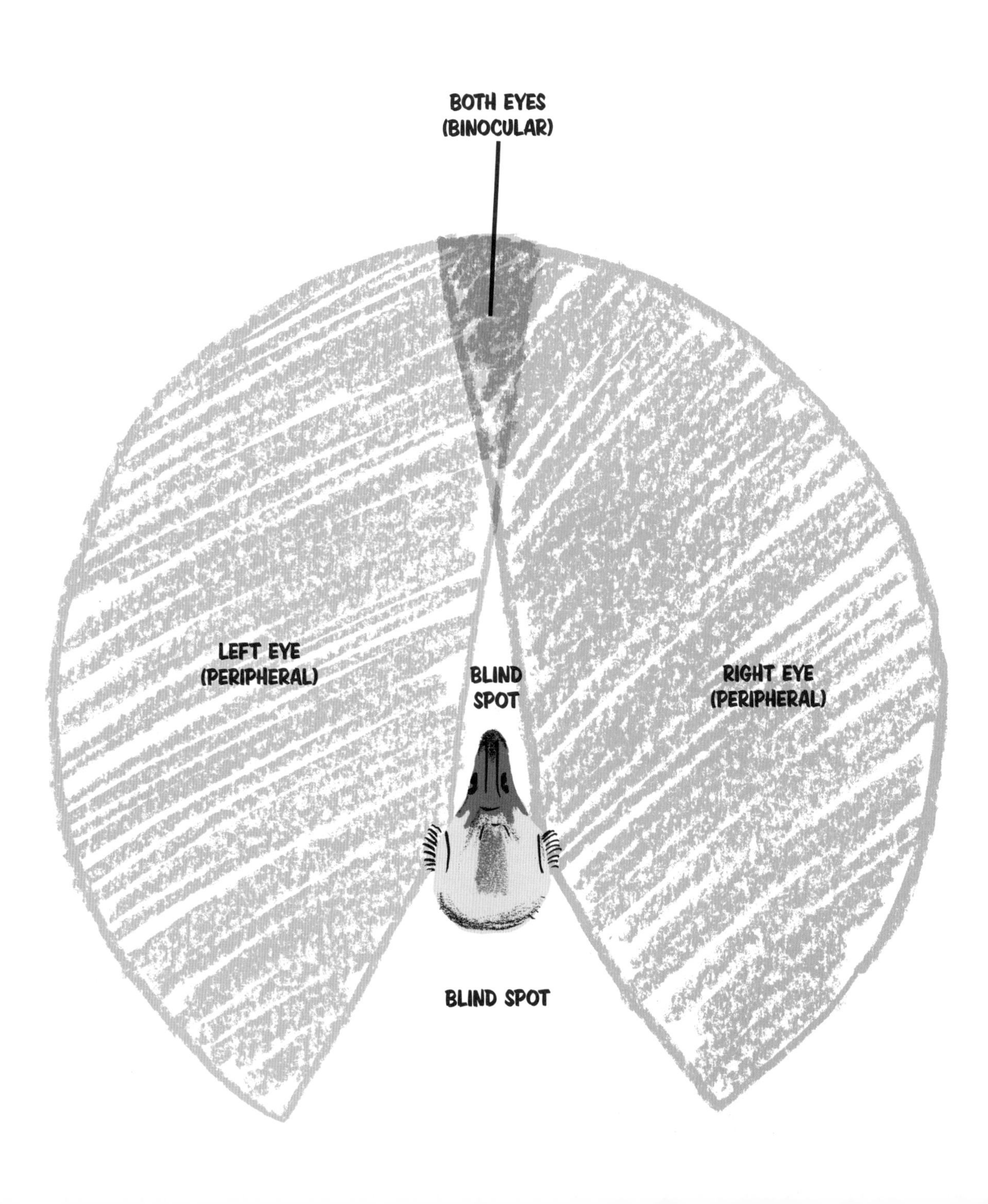

Predators like lions have eyes at the front of their head, which gives them wide ***binocular vision***—that's what both eyes see together. Binocular vision helps judge distance, so lions know exactly how far to lunge to get dinner. Prey animals often have eyes on the sides, which lets them scan more of the world for dangers.

And this is still just the beginning of what makes Eno's eyeballs amazing. Ostriches also have . . .

Look at it go! And look at the gazelles running. That cheetah only needed to twitch and the gazelles were off like lightning, darting and dashing in every direction.

PHEW! The gazelles escaped, this time. A few seconds of high-speed action, then back to the peaceful savanna. It's so hot here, nobody wants to run for long. But that cheetah looks hungry. I think we should go before the sun sets, don't you?

On the way home, how about a little quiz, just to pass the time? Don't worry—it's an **EYEBALL DOODLE QUIZ,** which means it is more doodle than quiz.

QUIZ #1

NOTE: If this is a library book, ***DON'T DRAW ON THIS PAGE!*** I don't want your teacher and your librarian angry at me.

1. ***EYEBALLS FOR BRAINS***

   Ostrich eyeballs are so big, they are bigger than an ostrich brain! Only slightly, but definitely bigger. Here's the size of your brain. And here's your head. Draw yourself eyeballs the size of your brain.

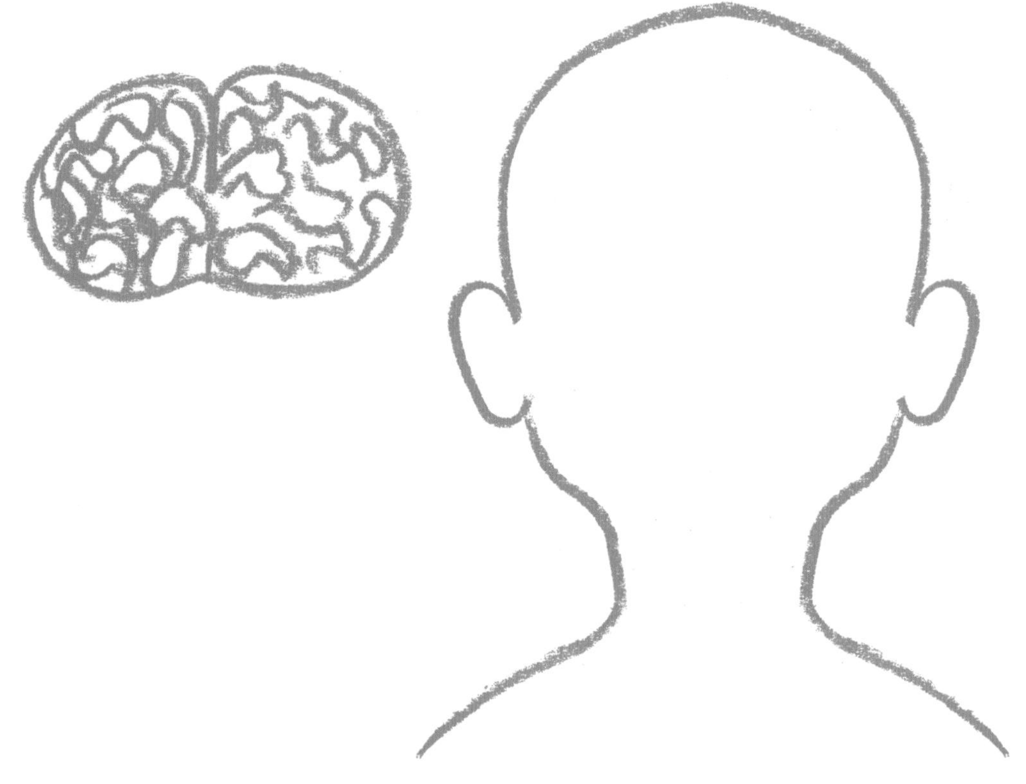

2. ***NO SQUINTING***

   Did you know birds have built-in sunglasses? Scientists think the pecten (that's the lump inside a bird eyeball) might also absorb glare, kind of like sunglasses inside your eyeball. And that might help birds spot predators or prey in dazzling light. Draw sunglasses on your giant eyeballs. While you're at it, give yourself a hat. Make it fancy!

3. ***SPOT THE LIONS***

Eno sees the world in way more detail than you do. Transfer the drawing in the large grid to the small grid to find out how many lions you would be able to spot.

ENO IS ONE OF A KIND . . . AND KIND OF WEIRD

THE SAVANNA HAS VARIOUS BIG CATS, like lions, cheetahs, and leopards. There are lots and lots of animals with hooves, like zebras, gazelles, giraffes, and wildebeests. There are doggy creatures, like African wild dogs and hyenas, which aren't dogs but sort of look like them. But there is only one giant bird on the savanna. Eno is one of a kind.

Another way of saying "one of a kind" is "kind of weird." Even I can't argue that ostriches are a strange combination of bits: supersize chicken body, snaky neck, bobble head, no-fly wings, horse legs, horrible toes—and those are just the parts you can see.

But "one of a kind" can also mean **"AMAZING! STUPENDOUS! WHIZ-BANG-WOW!"** And believe me, with all his weird bits working together, Eno is the fiercest, fastest warrior on the savanna.

There is a lot to explain, so let's hurry up and meet the family!

ENO'S FAMILY

HERE'S UMA. BUT LET'S NOT BOTHER HER—she's sitting on her eggs. She might have as many as twenty eggs under there. Her grayish-brownish feathers really blend in with the dry savanna grasses, don't they? Eno has other girlfriends (ostriches usually live in smallish herds of a dozen or less), and some of them have laid eggs in the nest, but Uma is his main squeeze. If any of those eggs hatch, it will be Eno and Uma who raise the chicks, no matter who laid them.

Ostriches belong to a strange group of flightless birds called ***ratites***. Most are big and fierce, but none are as big and fierce as ostriches. Rheas and emus look a bit like Uma, but smaller and shaggier, with stumpy legs. Cassowaries look like giant prehistoric turkeys with dinosaur-tooth hats. And then you have the kiwis, which are only chicken-size and very shy.

Not so long ago, much bigger ratites existed. New Zealand had giant moas that were thirteen feet (4 meters) tall! They were hunted to extinction shortly after humans first arrived on the islands, around 600 to 800 years ago. And Madagascar had elephant birds, the biggest birds ever. They were a little shorter than moas but heavier—they weighed as much as grizzly bears! They went extinct over two centuries ago, maybe because humans couldn't stop eating their eggs—each one was as large as 200 chicken eggs.

Since ratites live in Africa, South America, New Zealand, and Australia, scientists think their ancient relative flew around the world, then each family lost the ability to fly. As the birds got bigger, they were better able to defend themselves on the ground, but they grew too heavy for their wings to lift them.

And then, if you want to keep going back in time, we get to the dinosaurs.

Eno isn't directly descended from *Tyrannosaurus rex*—they are more like ancient cousins. But Eno, T. rex, *Velociraptor,* and all the fiercest meat-eaters of the dinosaur world are part of the same family tree. They are all ***theropods,*** which means "beast-footed." (See! I told you Eno had horrible feet.) Sure, they are millions and millions and *millions* of years apart, but Eno is way more closely related to T. rex than he is to you or a cow. To be honest, all birds are theropods, which means all birds are living dinosaurs of sorts. But of all the birds I know, none look more like dinosaurs than ostriches.

Birds and humans have had separate family trees for over 300 million years. That's way before dinosaurs even existed!

Next, I've told you Eno is big, but how big is he, ***REALLY?***

ENO IS REALLY BIG

YOU MIGHT KNOW OSTRICHES are the biggest birds alive, but did you know that male ostriches like Eno can be over nine feet (2.7 meters) tall? Enormous! And Eno also weighs a lot. Most birds fly, so they need to stay light. But not ostriches. Eno weighs close to 320 pounds (145 kilograms). Massive!

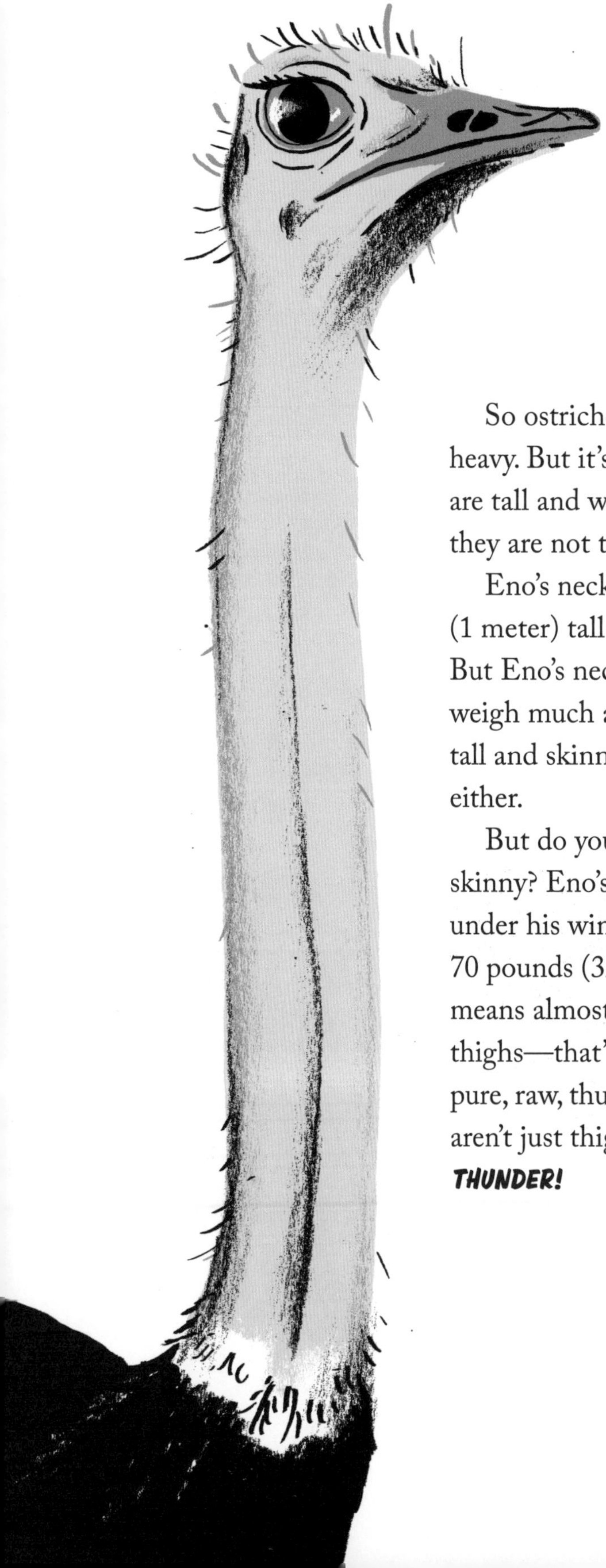

So ostriches are tall, and ostriches are heavy. But it's important to know which parts are tall and which parts are heavy, because they are not the same.

Eno's neck is almost three and a half feet (1 meter) tall. Only giraffes have taller necks! But Eno's neck is super skinny and doesn't weigh much at all. Eno's legs are also super tall and skinny—they don't weigh much either.

But do you know what isn't super tall and skinny? Eno's thighs. It's a bit hard to see under his wings, but each thigh weighs about 70 pounds (32 kilograms). Times two! Which means almost half of Eno's weight is in his thighs—that's 140 pounds (64 kilograms) of pure, raw, thundering thigh muscles! Those aren't just thighs. Those are ***THIGHS OF THUNDER!***

SUPERPOWER #2

THIGHS OF THUNDER

BEFORE WE TALK ABOUT ENO'S THIGHS, I need to explain a few things about his legs. Humans and ostriches both walk on two legs. (Scientists call us *bipedal.*) And we both have thighs, shins, feet, and toes, but Eno's legs are nothing like yours. It's like he has a whole extra leg joint. I'll need a diagram to explain.

- **THIGH BONE:** Eno's thigh bones don't hang vertically below his body, like yours. They're stuck on his sides and almost horizontal.

- **KNEE:** Imagine having knees at your rib cage.

- **SHIN BONE:** What looks like Eno's thigh is actually his shin.

- **ANKLE:** What looks like a backwards knee is actually the same joint as your ankle. Weird, right?

- **FOOT BONE:** What's even weirder is Eno's enormous foot bone. It's made from the same bones in your ankle and the sole of your foot all fused into one giant bird bone called a *tarsometatarsus.* A big word for a big bone, huh?

- **TOE JOINT:** What looks like Eno's ankle is the same joint as your toe-flexing joint.

Okay. Now back to Eno's thighs!

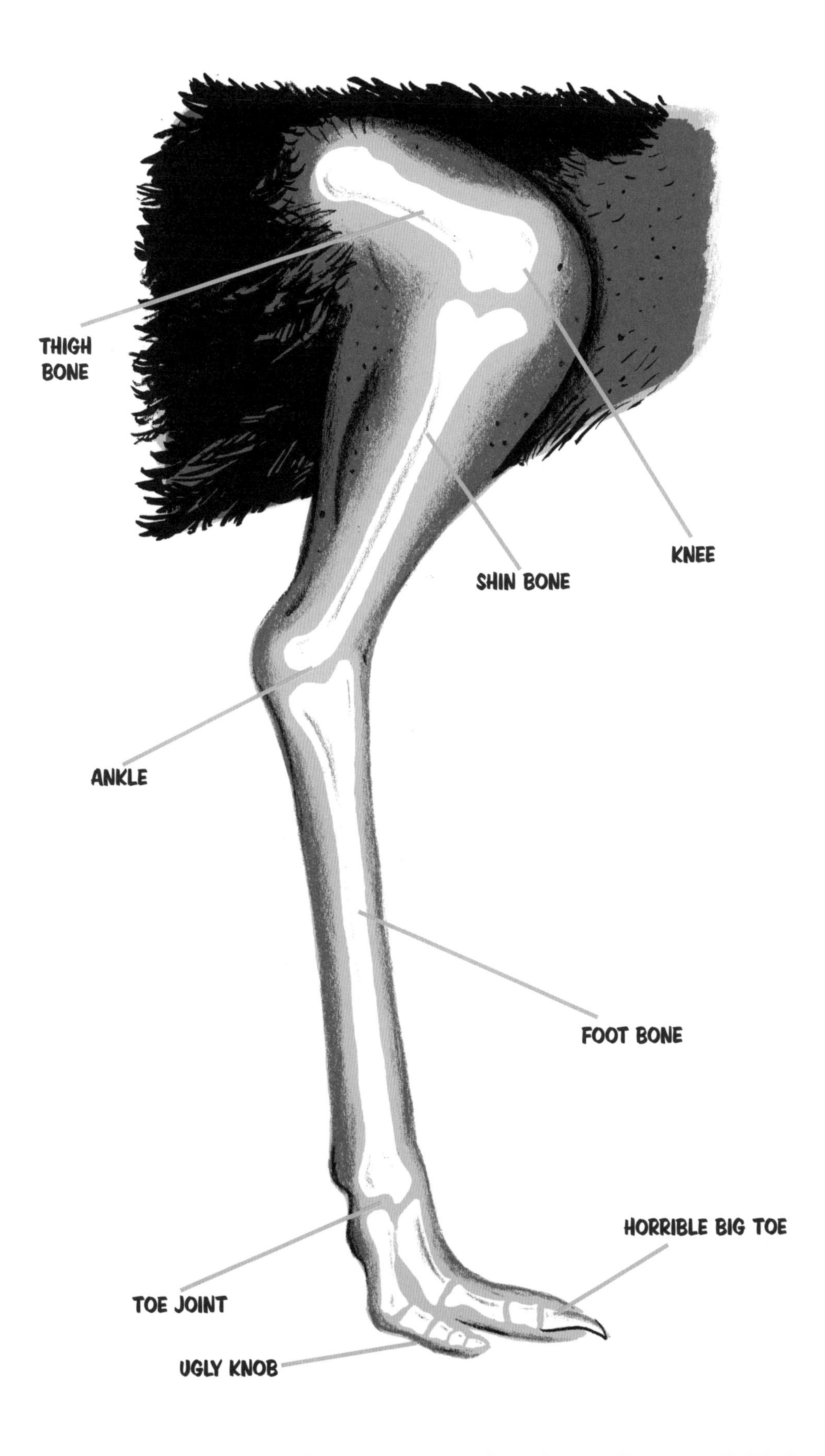
THIGH BONE
KNEE
SHIN BONE
ANKLE
FOOT BONE
HORRIBLE BIG TOE
TOE JOINT
UGLY KNOB

You may already know that muscle is the meaty stuff that moves our bodies. Bigger muscles are stronger muscles. Bigger muscles can lift more weight and push your legs faster to win the race. With his ***THIGHS OF THUNDER,*** Eno can win most races. (His skinny legs help too, but I'll explain that in a bit.)

Now, ostriches are the fastest two-legged animals on earth, which sounds impressive, but isn't. Only humans and birds really run on two legs, and ostriches are much, much faster than humans. Ostriches can run forty miles per hour (64 kph). The world's fastest human can only run about half as fast. And let's be honest, most of us can't run nearly as fast as the world's fastest human.

What's much more impressive is ostriches are faster than most four-legged animals as well. Eno is faster than your dog. He's faster than a horse, except maybe a prize-winning racehorse. He's faster than a hyena. But—and this is important—Eno is not faster than a cheetah. Cheetahs can reach seventy-five miles per hour (121 kph) and are the fastest sprinters in the world.

Luckily, cheetahs usually hunt alone, and Eno is too fierce for a lone cheetah. But if a cheetah ever did try to run Eno down, it could get a nasty surprise. You see, Eno's thighs can do more than run. When under attack, those mighty thighs pack a killer punch.

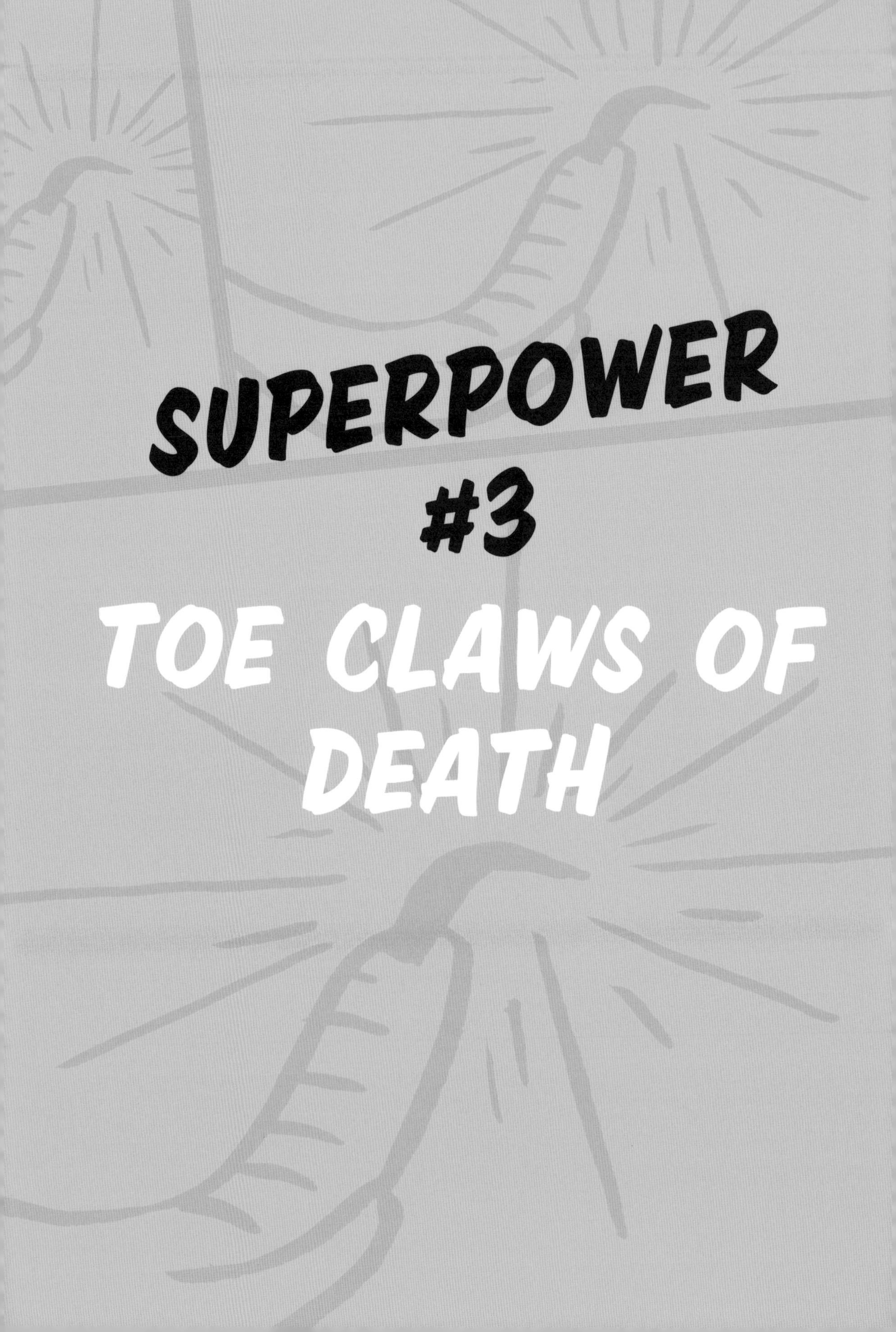

SUPERPOWER #3

TOE CLAWS OF DEATH

OSTRICHES TRY TO RUN FROM PREDATORS. But sometimes there is no choice but to **FIGHT!** An angry ostrich will fluff out its enormous wing feathers. It will hiss and lunge and menace. And when the time is right, it will kick with all the power in those **THIGHS OF THUNDER.**

Most birds have hollow bones, which make them lighter for flying. Ostriches have hollow bones too, but they're gigantic and heavy—as big as a horse's or a cow's leg! A direct kick from an ostrich can break a lion's ribs. It can knock the fight right out a cheetah and send it flying. And once it kicks that big cat down, an ostrich will stomp on it. If the lion's ribs weren't already broken, they're broken now. And that's not all an ostrich kick can do.

Remember Eno's toenails? They're black. They're nasty. And when they are on the kicking end of Eno's thundering thigh muscle, they become ***TOE CLAWS OF DEATH!***

Ostriches kick out in front, then rip downward with their dagger-like toe claw. A slash from an ostrich claw can disembowel a lion. (Do you know what *disembowel* means? It means the lion's guts aren't in its body anymore, which is a very, very bad thing for the lion.)

And ostriches don't just kick lions. They'll kick anyone or anything that makes them angry, so don't get too close. I mean, you wouldn't get too close to a T. rex, would you?

Because of the way ostrich legs are built, ostriches can't kick backwards! They can only kick forward and about chest high. Someone told me the best way to avoid an ostrich attack is to lie flat on the ground. They might still stomp on you, but you'll be saved from that ***TOE CLAW OF DEATH.***

DON'T TELL ENO

I NEED TO TELL YOU A SECRET. I'm whispering, so Eno doesn't hear.

Eno lives in Africa, where ostriches have lived for millions of years. Yes, life is tough. And yes, a lion might attack at any moment, but Eno is free to roam and peck and run, just as he pleases.

However, most ostriches don't live in Africa. Most live somewhere else around the world on farms. Australia, Canada, Norway, Thailand . . . you name a country and I bet it has an ostrich farm. Why? Because people like to eat ostrich thighs. (That's why I'm whispering.)

And before farmers raised ostriches for meat, ostriches were hunted for their feathers. Kings and emperors, queens and elegant ladies thought ostrich feathers were so glamorous they put them in their hair, on their hats and headpieces, on their helmets, even on their horses. In fact, people loved ostrich feathers so much, they almost hunted the giant birds into extinction.

Ostrich feathers are still used in fashion sometimes and as feather dusters. But these feathers are from farmed ostriches. Today, more ostriches live on farms than in the wild! In fact, the idea to farm ostriches for meat and leather probably saved ostriches from extinction.

But promise you won't tell Eno any of this, okay? Now back to Eno's legs. They may be super skinny, but those legs are supersonic!

SUPERPOWER #4

SUPER-FANTASTIC ELASTIC STRIDERS

ON THE WIDE-OPEN PLAINS OF THE SAVANNA, where predators can always see lunch, and lunch is always on the lookout, you have to be fast to survive. It's not surprising that some of the world's fastest animals live here.

To run faster, you can do two things. You can lengthen your stride or you can speed up how fast your legs make that stride. Of course, Eno does both.

Eno's **THIGHS OF THUNDER** are the power in his engine, but if Eno had short and stumpy legs he'd have a short and stumpy stride. Think of turtles' legs . . . not fast at all. But Eno doesn't have stumpy legs. Eno has ridiculously long legs, which means he has a ridiculously long stride. Eno might cover as much as 11.5 feet (3.5 meters) with every step!

And there's more. Eno's muscly thighs are high on his body, but his lower legs are much skinnier. Skinny means super lightweight, and that means Eno can drive his legs further and faster with much less energy.

11.5-FOOT (3.5-METER) STRIDE

And that's still not all! Those skinny legs also have their own **SUPER-FANTASTIC ELASTIC POWER** built right in. You see, Eno's long legs have superlong tendons. ***Tendons*** are like the body's elastic bands, holding everything together. And Eno's superlong elastic bands work like a bow and arrow. Every time Eno pushes down on a leg, his joints bend and the tendons store up energy, like a drawn bow. Then, with every stride, that stored elastic energy shoots Eno forward. This lets Eno use 50 percent less thigh-muscle power than you would, which means his muscles won't tire out as quickly. Talk about having a spring in your step!

It's a bird . . . it's an arrow . . . it's the **AVIAN ROCKET!** Do you know the word ***avian?*** It's just a fancy word for bird. But "bird rocket" lacks pizzazz, don't you think?

The spring also gives Eno a smooth and effortless stride—he looks like a super-graceful, supersonic, superfierce ballet dancer.

But wait! Those legs still have one more secret to their super speed.

SUPERPOWER #5

TWO-TOED TORPEDOES

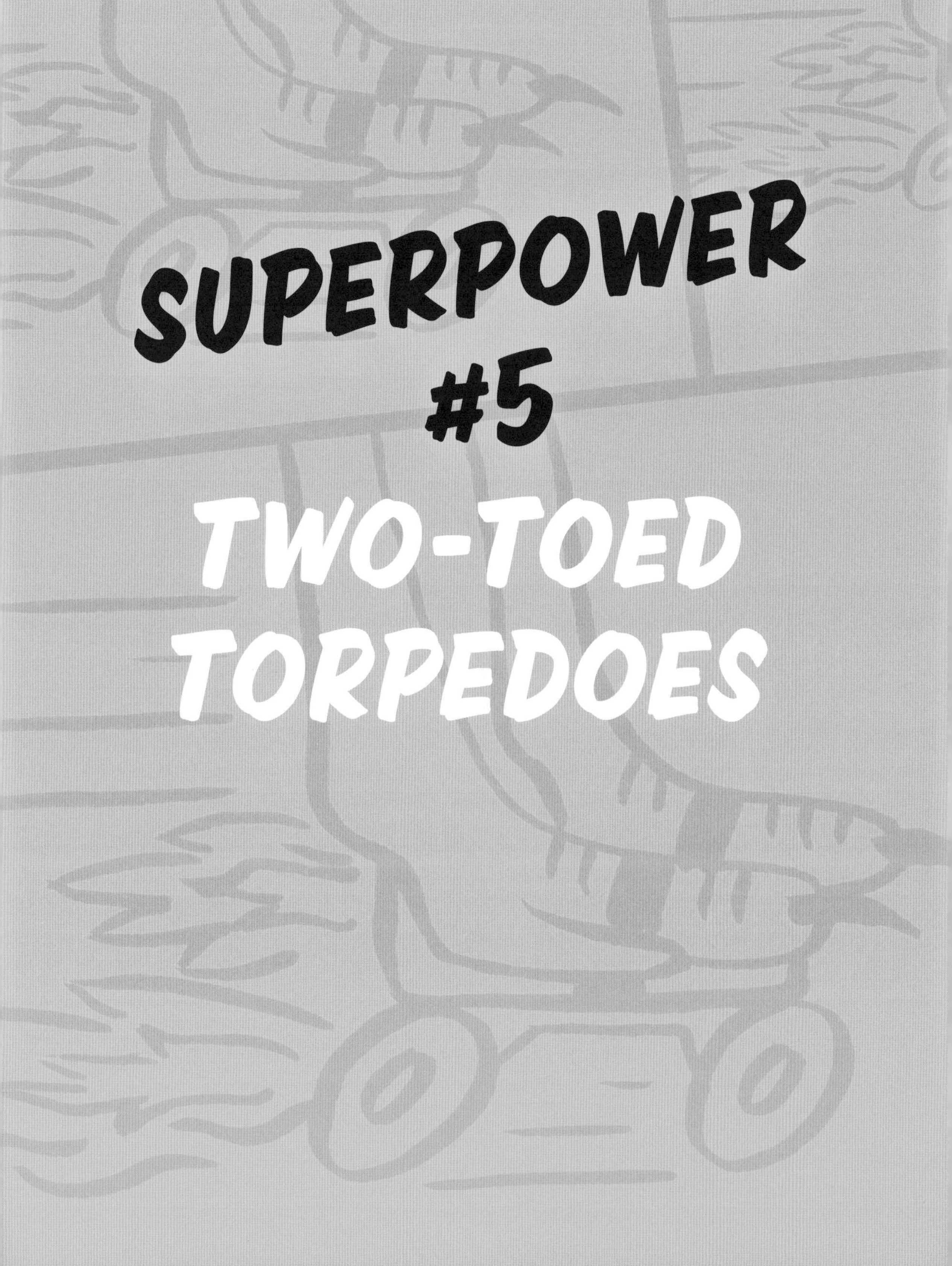

OF ALL THE ANIMAL LEGS AND FEET IN THE WORLD, which do ostriches have the most in common with?

Did you guess cheetahs? Nope. Not hyenas either. And no, not even other birds.

UNGULATES! ***Ungulates*** are animals with hooves, like gazelles and horses. As you may know, gazelles and horses are some of the fastest animals around. And, like ostriches, they get their super speed from long, skinny legs powered by big thigh muscles high on their bodies.

But ungulates and ostriches have also both lost toes. Horses' ancient ancestors had five toes, just like you. But over millions of years, their middle toe grew longer and longer while their other toes disappeared. The reason is simple: fewer toes mean lighter legs, which can move faster with less effort. The same goes for ostriches. Losing toes has turned Eno's feet into **TWO-TOED TORPEDOES!**

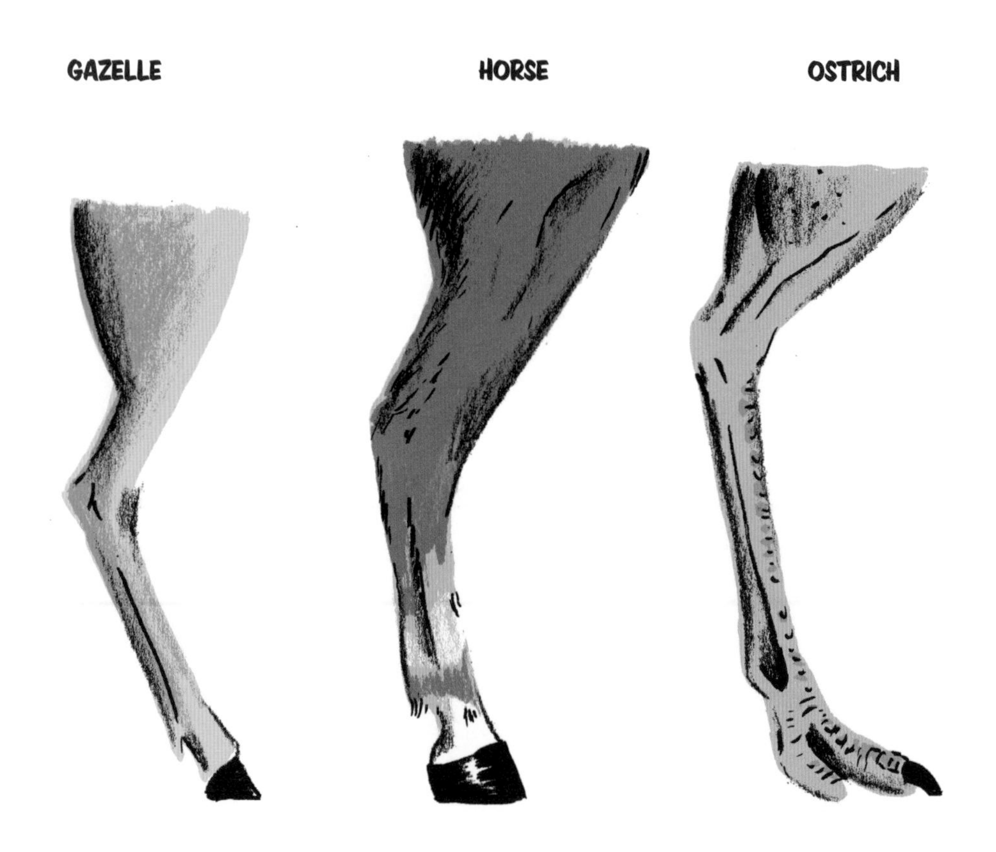

Ostriches are the only birds with two toes. (Most birds have three or four.) Ostriches also walk and run on their tippy toes. (Most birds walk flat-footed, like ducks.) It saves time and energy to never put your whole foot down, so Eno's ugly toe knob only touches down to keep balance. Also, having fewer bones means Eno's legs are extremely stable during high-speed running.

And don't forget those **TOE CLAWS OF DEATH!** With every stride, Eno's thundering thighs hammer those claws into the ground for extra grip.

SPEED! POWER! GRACE! DEATH KICKS!

Okay. Time to put your speedy knowledge to the test. But take a deep breath because **QUIZ #2** is the fastest quiz around!

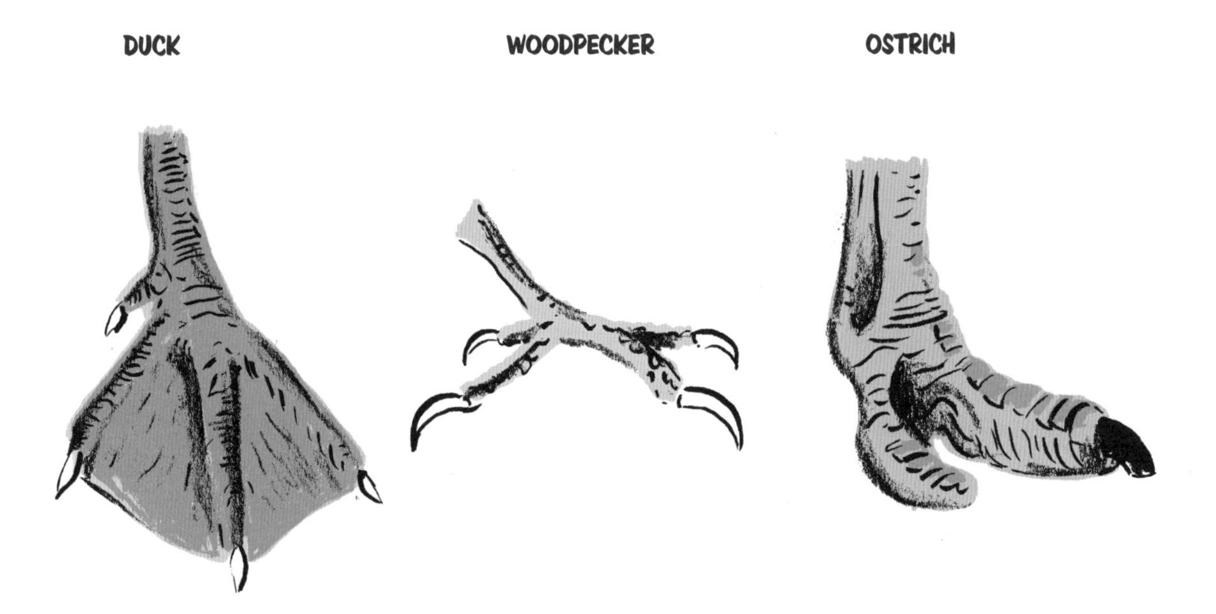

*NOT TO SCALE

QUIZ #2

PLEASE CHOOSE THE CORRECT ANSWER FROM THE CHOICES BELOW.

1. Who would win a 100-meter dash?
   - **a.** you
   - **b.** T. rex
   - **c.** Eno
   - **d.** kiwi bird

2. You know Eno lives in Tanzania, on the east coast of Africa. Let's say you live in New York City. The fastest way to travel from New York City to Tanzania is by
   - **a.** rowboat.
   - **b.** ostrich.
   - **c.** motorcycle.
   - **d.** It's a trick question: motorcycles and ostriches are the same speed.

3. Ostrich eyeballs are as big as cue balls, but they weigh almost four times less. That means each of Eno's enormous eyeballs weighs about as much as a(n)
   - **a.** mouse.
   - **b.** match.
   - **c.** lemon.
   - **d.** egg.

4. If you lived in Florida in 1902 and someone said, "hitch up the ostrich," you would
   - **a.** hitch up an ostrich to a small carriage.
   - **b.** pull up your feathered socks.
   - **c.** open your umbrella.
   - **d.** grab a friend and do the Ostrich polka.

5. Humans have 62 bones in their legs and feet. How many bones does an ostrich have in its legs and feet?
   - **a.** 348
   - **b.** 2
   - **c.** 47
   - **d.** 30

ANSWERS:

1. **c.** Ostriches every time. T. rex was huge but its top speed was around 15 mph (24 kph), which is probably faster than you can run but slower than an Olympic sprinter and most definitely slower than an ostrich! As for a kiwi bird, they have a superfast sprint, but they wouldn't do very well in a race. They're shy, skittish, and almost blind, and would probably just hide in the nearest bush until the race was over.

2. **a.** This is a ridiculous question because there is a giant ocean between New York City and Tanzania. Ostriches can't swim. Neither can motorcycles. So your only chance of getting there is a rowboat. It will take you about a year, and you'll probably capsize, but at least you have a chance. And in case you're interested, ostriches might be superfast, but a motorcycle is still way faster.

3. **d.** A medium egg weighs 1.7 ounces (50 grams), or about as much as one of Eno's eyeballs.

4. **a.** At the turn of the century, Florida was a hotspot for ostrich racing. A driver would hitch an ostrich to a small carriage, and away they'd go! Around a track or maybe just down the road. At the Florida Ostrich Farm in Jacksonville, a young ostrich named Oliver W. was famous for being the fastest animal around. Apparently, he loved his carriage and could run a mile in two minutes and twenty-two seconds. His owners said he was better than a horse because he ate less, never shied or ran away, and went a good pace without laziness or fatigue.

5. **d.** Humans have 10 leg bones (including two kneecaps, one on either side), 14 ankle bones, and 38 foot bones. Ostriches have 12 leg bones (including four kneecaps, two on either side) and 18 toe bones. In other words, humans have more bones in their feet than an ostrich has in its entire leg! And that's including all those kneecaps. ***BONUS FACT:*** Ostriches are the only animals with four kneecaps. Wonders upon ostrich wonders!

DINOSAUR WINGS

I KNOW THIS IS A BOOK ABOUT OSTRICHES and not dinosaurs, but it's really hard to talk about ostriches without talking just a bit about dinosaurs. Even paleontologists (those are dinosaur scientists) are interested in ostriches. Some study ostriches' long, hose-like necks to find clues on how long-necked dinosaurs worked. Others study ostrich legs to learn how long-legged raptors might have ran. And then there are Eno's wings.

You know Eno can't fly. But if you think Eno's wings are useless no-fly flappers, you would be both right and wrong. You see, ostriches have wings because their ancient bird relatives flew, so ostrich wings are ancient leftovers. But that does not mean ostrich wings aren't hardworking. Eno's no-fly flappers might actually be prehistoric **DINOSAUR WINGS!**

Have you seen pictures of *Gigantoraptor*? He was an enormous bird-like dinosaur with little feathered wings. *Gigantoraptor* most definitely did not fly, but those little wings helped him run like the wind. Like flaps on a plane's wings, *Gigantoraptor*'s feathered arms caught the air to give him extra lift. In fact, scientists now think feathered wings were originally used by dinosaurs for lift and balance way before any birds or dinosaurs learned to fly.

Now, dinosaur wings certainly didn't help all the dinosaurs survive, but in the scorched savanna, surrounded by prowling lions, wings definitely give Eno a **DO-IT-ALL SURVIVAL ADVANTAGE.**

SUPERPOWER #6

DO-IT-ALL DINO FLAPS

ENO'S WING BONES ARE PRETTY SMALL for his size. But his feathers are enormous. When Eno lets his wings down, the tips touch the ground. Ostrich feathers are soft and downy, not like the stiff flying feathers on a crow or seagull. In fact, they are so soft and fluffy, they keep ostriches warm on cold nights. But you know chilly weather isn't Eno's biggest problem.

What is his biggest problem?

Actually, Eno has three: not getting eaten, beating the heat, and finding a girlfriend. And here are four ways his **DO-IT-ALL DINO FLAPS** get the job done.

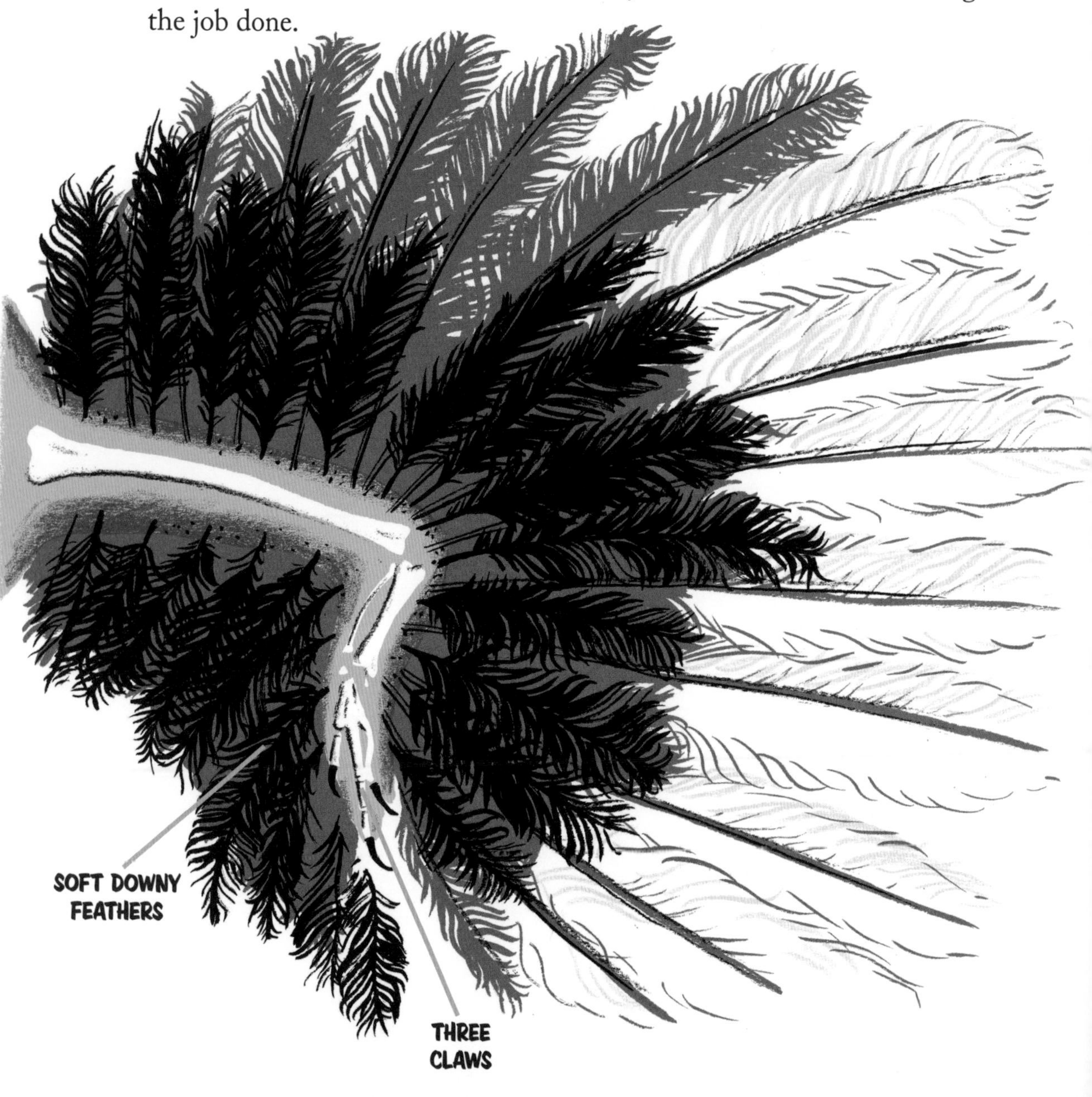

FIRST: They are rudders. Eno uses his wings to get extra lift while running at top speeds. They also help him make quick turns and fancy zigzag moves without falling over.

SECOND: They are air breaks for fast stops. He also fans out his wings to look bigger and meaner when facing down a lion.

THIRD: They help him cool off. Eno's mighty thighs are naked under his wings. In hot weather, Eno lifts his wings to let the breeze blow on his bare skin. Aaahhh . . . so pleasant.

FOURTH: They are super pretty. Some boys buy flowers to impress the girls. Eno does a ridiculous feather dance. He bows and flip-flaps. He fluffs and preens and fancy-steps. Plus, the skin on his neck and legs turns bright red. And he can puff out his neck skin to make deep hooting noises to scare off other boys. I think Uma was impressed.

BONUS: Ostrich wings have claws! If you need proof that Eno is T. rex's little cousin, just look at those three awful claws on each wing. They're leftovers from ancient relatives that had clawed hands.

BREAKNECK SPEED

ENO'S LEGS HAVE MORE HIGH-SPEED POWERS than your dog can count, and his wings help out too. But there is still more to Eno's super speed. You see, Eno is built for **SUPERSONIC SURVIVAL,** where life or death is a matter of inches or seconds away. It isn't simply that Eno has enormous eyes to spot danger or supersonic legs for escaping—every single bit of Eno is perfectly built so his eyes can be bigger than big and his legs faster than fast.

Take Eno's neck, for example. That wiggly hose is perfectly designed for high-speed running **AND** high-powered looking. Let me explain.

You might know that giraffes have extra-tall necks to reach the high leaves of acacia trees. But ostriches mostly eat things off the ground, like seeds, leaves, roots, bugs, and maybe a lizard or two. They also swallow stones and grit for grinding food in their ***gizzards*** and drink water from the watering hole. And all of those things happen on the ground.

But that's precisely why ostriches need such long necks. If we gave Eno a shorter neck—maybe half as long—his head wouldn't reach the ground. He would either have to kneel down for every bite (absurd!) or his legs would also have to shrink by half. But if Eno's legs shrunk by half, he might only run half as fast, which means he probably couldn't outrun that hungry lion.

It was long believed that ostriches could digest metal (they can't), probably because ostriches like shiny things and will eat almost anything. In 1930, the stomach of a dead ostrich was discovered to contain dozens of nails, hooks and other metal bits, seven coins, a pencil, a key, a bit of wood, three gloves, three handkerchiefs, and a long stretch of rope.

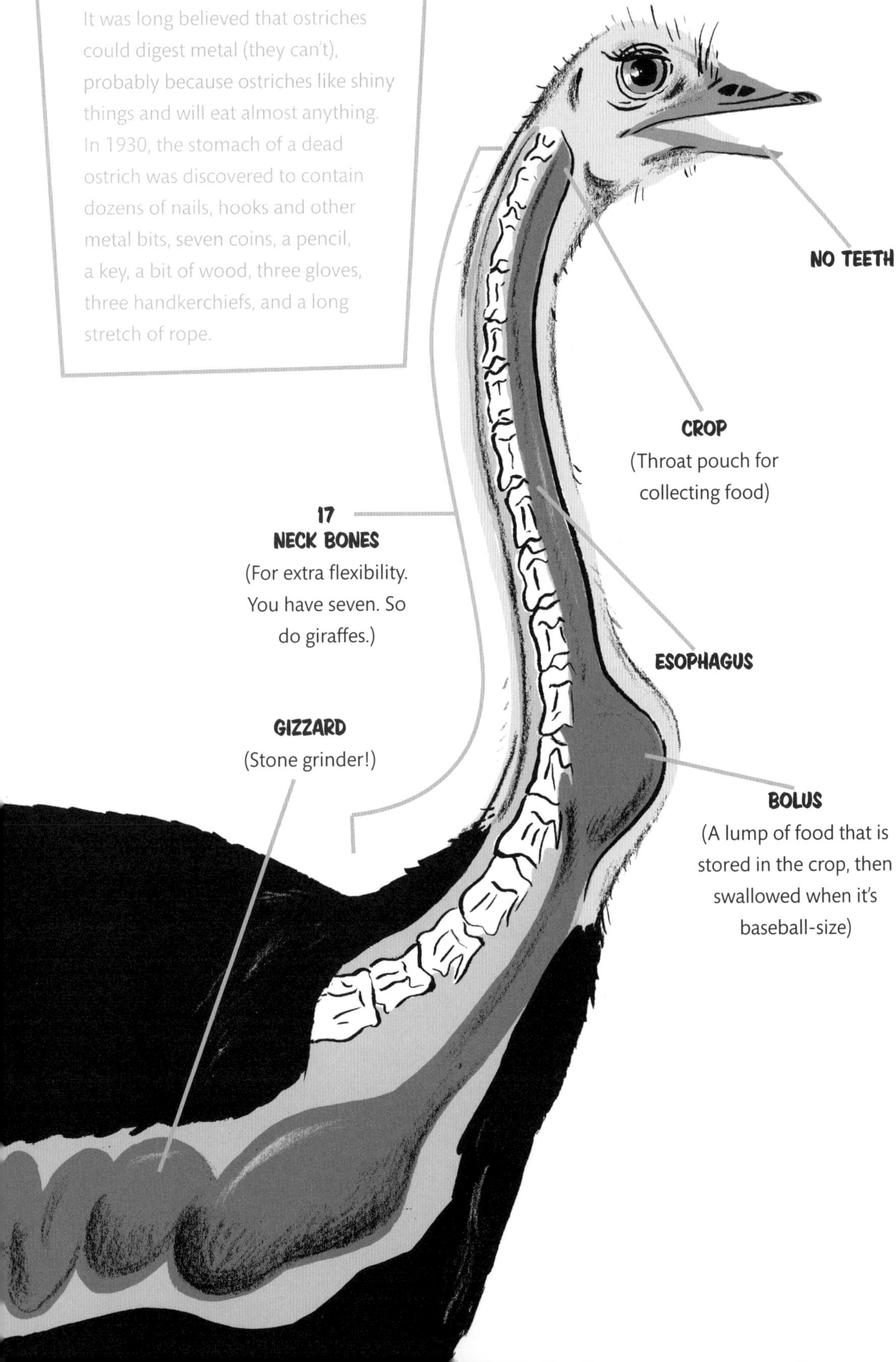

NO TEETH

CROP
(Throat pouch for collecting food)

17 NECK BONES
(For extra flexibility. You have seven. So do giraffes.)

ESOPHAGUS

GIZZARD
(Stone grinder!)

BOLUS
(A lump of food that is stored in the crop, then swallowed when it's baseball-size)

In other words, Eno is a seesaw survival puzzle. The longer his legs, the faster he can run. But longer legs mean Eno's head is further from the ground, which means his neck needs to be super long to reach food and water. So, you see, it is true—Eno's neck does help him run fast!

And that neck also puts the **TELESCOPIC POWER** in Eno's **COLOSSAL ORBS OF TELESCOPIC VISION.** Even the world's best binoculars can't see far if you're lying on the ground. But climb a tree and you can see for miles. Same goes for ostrich necks. The longer the neck, the higher the eyeballs, the further the ostrich can see. Having such a long, flexible neck also lets Eno twist his head to see behind him or preen the feathers on his back. What an amazing neck!

But there is also some very bad news about having such a long neck.

You know superheroes have all sorts of different superpowers. Wonder Woman has super speed. Black Panther can walk on water. The Invisible Man is, well . . . invisible. But not all superheroes are super at everything. Even Superman has his weakness with Kryptonite. Eno has a weakness too, although it's more of a downside, or maybe a dark side, to his super speed. You see, Eno's weakness is his tiny head.

NOT A SUPERPOWER: TINY HEAD

THE PROBLEM IS SIMPLE: ostrich necks are really more like garden hoses than necks. And if there is one thing I know about garden hoses, it's that they are very long, very bendy, and not very strong. A garden hose could never hold up anything heavy. It could hold up an egg. Or maybe a tennis ball. But that's about it.

Same goes for Eno's neck. It's way too long and wiggly to hold up a heavy head. Let's be honest, Eno's head is tiny. In fact, Eno's head is so tiny, there is only room for two very large eyeballs and not much else. What I'm trying to say is that inside Eno's tiny head is a tiny brain. (I already told you Eno's eyeballs are bigger than his brain, right?)

Now, lots of birds have tiny heads and are amazingly smart. Crows can count and solve puzzles. Parrots can use tools and talk. But Eno is not one of those birds. Eno's tiny brain is just tiny.

A tiny brain can get ostriches into trouble. Ostriches could probably outpace most lions, if they ran in a straight line. But ostriches often run in circles. Don't ask me why.

A tiny brain is always a bummer. I'm not going to say it isn't. But that tiny brain is part of the whole seesaw survival puzzle that lets Eno's superpowers shine. You see, a tiny head lets Eno have a long and wiggly neck, which means he can have superlong legs for high-speed getaways. So in a sense, Eno's tiny brain lets him run faster to survive another day.

I'm sure ancient ostriches could have evolved smaller eyeballs to allow room for bigger brains. But what do you think is more important in the Serengeti: solving math problems or seeing hungry lions with your ***COLOSSAL ORBS OF TELESCOPIC VISION?*** I vote for colossal orbs. So did ostriches.

Of course, that's not the way nature works. Ostriches didn't vote for long legs or giant eyeballs. It just happened very, very slowly, over millions of years. And it happened because longer, faster legs and bigger, better eyeballs helped ostriches survive in a wild world.

You see, no two living things are exactly the same, even though they might look mostly the same. One ostrich might have slightly longer legs than another. The ostrich with slightly longer legs might have a slightly better chance of outrunning a lion, which means it has a slightly better chance of surviving long enough to have slightly longer-legged babies. And so on, and so on, over millions of years.

Scientists call this "survival of the fittest" or "natural selection," which are very clean and tidy words for the wild, weird, and downright wacky ways all the creaturely life that ever lived on earth survived and slowly, slowly, over billions of years, changed from floating blobs in ancient seas to become all the beasts and birds and bugs we know today.

The very big word for this is ***EVOLUTION***. In fact, evolution is such a big word it needs its own book. But I don't have time for that right now. I want to tell you about rocket fuel instead. And lungs. And strange, impossible superpowers. Sound good?

THE PUFF AND SQUASH

HAVE YOU READ MY BOOK about Rosalie the Mole, ***BIONIC BURROWER?*** If you haven't, you should. But if you have, you'll know a bit about how lungs work.

Moles are mammals, just like us, and our lungs work pretty much the same. As we suck air in, our lungs puff up like two balloons. These balloons are actually made of millions of the tiniest, itsy-bitsiest balloons called ***alveoli***. Each lung has about 250 million—that's 500 million in total!

As we breathe in, each of our 500 million tiny balloons fills with air. Then, our blood flows alongside every tiny balloon and grabs the important stuff—called ***oxygen***—from the air inside. Oxygen is the reason we breathe in at all; it keeps us alive and our bodies working strong. Our blood then rushes around, delivering fresh oxygen to every bit and corner of our whole body. So that's **STEP ONE: THE PUFF.**

Then comes **STEP TWO: THE SQUASH.** Our lungs deflate and squash all the old, stale air out. Then back to breathing in again. In, out. In, out. Puff, squash. Puff, squash.

OUR LUNGS

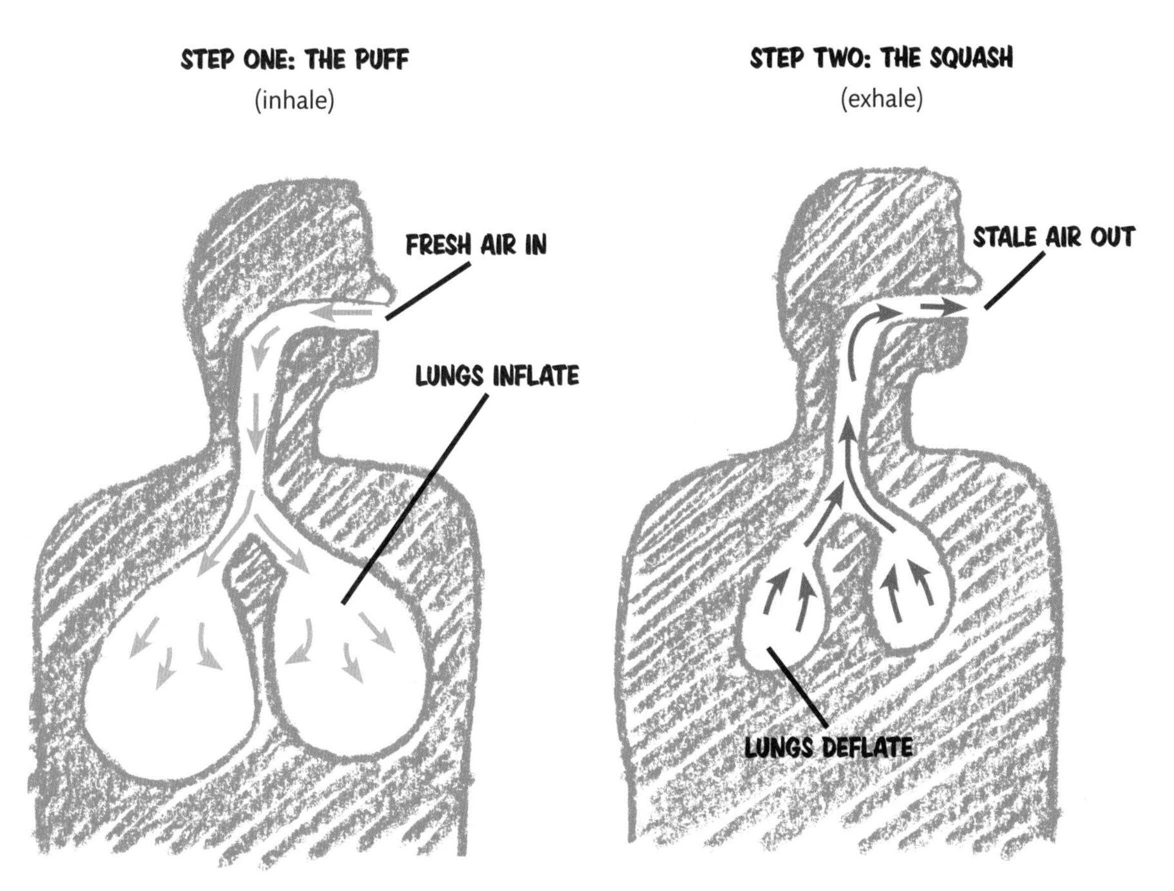

(So I don't get in trouble with your teacher again, the proper word for ***PUFF*** is inhale. The proper word for ***SQUASH*** is exhale.)

So far so good? Good. Because that's not the way bird lungs work. At all. Of course, birds still need to get oxygen in and stale air out, but everything else is different. And their lungs aren't just different than ours. They are much, much better. Impossibly, amazingly, almost magically better.

ALL PUFF, NO SQUASH

LOTS OF SUPERHEROES HAVE AMAZING LUNGS. Take Superman: He can blow air with hurricane force. He once sucked an entire tornado into his lungs and blew the twister into space. His freezing breath can turn anything into a block of ice. But even Superman needs to breathe in and out, just like you.

But not Eno! His lungs do the impossible. It's as if his lungs—are you ready?—it's as if his lungs never, ever breathe out! Impossible! Amazing!

You look puzzled. You're thinking, "What's so impossibly amazing about never breathing out?"

Well, to explain the ***IMPOSSIBLE AMAZINGNESS OF THE AVIAN LUNG,*** I, your Brave and Noble Guide to the Superpowered Heroes of the Animal Kingdom, will take off my jacket, tighten my shoelaces, and challenge Eno to a race.

I'm sprinting as hard as I can! As I run harder and faster, I need to breathe harder and faster because my muscles need more and more oxygen to keep working strong. But soon I just won't be able to breathe in fast enough . . . *puff* . . . because I have to breathe out the old air . . . *squash* . . . before I can breathe in . . . *puff* . . . more fresh air . . . PHEW! I give up. Eno is out of sight anyway.

You could say, and I certainly do, that breathing out is a big, fat waste of time.

MORE PUFF PLEASE!
LESS SQUASH!

Imagine if I could breathe in and in and in and in and in. Of course, my lungs would pop like overblown balloons. But if for some mysterious reason they didn't pop, I'd have **ROCKET LUNGS!** With no squash, an ever-flowing flood of fresh oxygen would endlessly stream through my lungs. ***NO SQUASH MEANS SUPERPUFF!*** With all that extra oxygen, my muscles could work a whole lot harder, and I could run much harder for much, much longer before I got winded. I might never stop!

BUT (and it's a big but) even if my lungs didn't pop, if I never breathed out, I'd poison my blood and die. You see, as blood races around delivering oxygen, it's also picking up our body's garbage gas and carrying it back to our lungs to breathe out. When your muscles gobble up oxygen, they burp out something called ***carbon dioxide.*** Think of carbon dioxide as the fumes belching from a race car. You and that car need to get rid of your fumes or your engines will sputter, fizzle, and die. ***NO SQUASH MEANS KAPUT!***

So, now do you see why bird lungs are so supernaturally superpowered? It's as if they can endlessly breathe in and in and in and still belch out their garbage gas. ***SUPERPUFF AND NO KAPUT!***

But no one can breathe in and out at the same time. Surely that's impossible . . . right?! Yes, it's impossible, for you. But not for Eno.

So can you guess how he does it? A hidden exhaust pipe? A secret second mouth? Nope and nope. Eno still breathes in and out through his mouth, just like you.

IT'S LIKE MAGIC! It really is. As with all magic tricks, it isn't really magic, just a really great trick. But that doesn't make the brain-boggling biological impossibility any less brain-bogglingly amazing. So sit back, relax, and allow me to explain the superpowered mysteries of ***THE IMPOSSIBLE EVER-FLOW LUNG!***

MORE PUFF =
MORE OXYGEN
=
MUSCLES
WORK
HARDER!

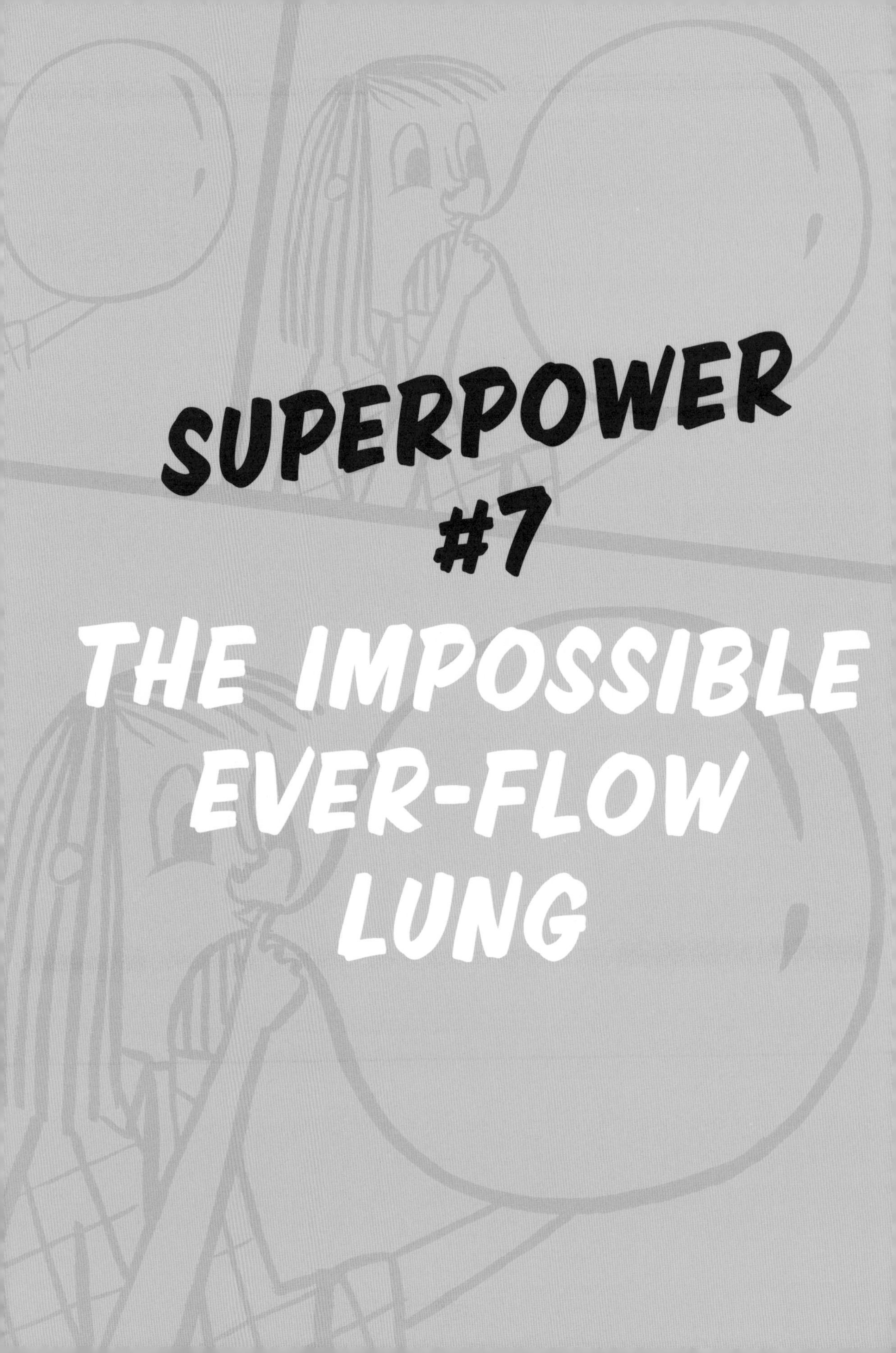

SUPERPOWER #7

THE IMPOSSIBLE EVER-FLOW LUNG

FIRST, BIRD LUNGS AREN'T BALLOONS. They are stiff tubes. The fancy word for these tubes is ***parabronchi***, but I'll just call them tubes. At either end, these tubes are connected to balloons—seven to twelve in total, depending on the bird. Ostriches have ten. These balloons take up about a fifth of the space in a bird's body—that's a lot! They squeeze around a bird's organs; some are even inside its hollow bones.

Now, bird balloons are part of the whole lung system, but they are not actually lungs. And they are not made from millions of tiny alveoli like your lungs. They are just basic balloons. They all have names, but I'll keep it simple. I'll divide them into two balloon teams: ***TEAM FRESH*** and ***TEAM STALE***.

Every time Eno breathes in and out, all the balloons on both teams puff and squash at the same time, just like your lungs. But the trick is having the two teams doing different things at different ends of the lungs: Team Fresh only deals with fresh air going in and Team Stale only deals with stale air going out. That, my friend, is the whole secret of bird lungs. Let's take a closer look.

> Scientists call Team Fresh, the posterior air sacs. *Posterior* is a fancy word for "bum" and *air sac* is a not-so-fancy word for "balloon," so the posterior air sacs are the balloons closer to a bird's back end. Scientists call Team Stale the *anterior* air sacs. They're closer to a bird's head.

STEP ONE: PUFF #1

Eno breathes fresh air into Team Fresh only. This is just like your puff, except bird balloons are not actually lungs. They're just balloons.

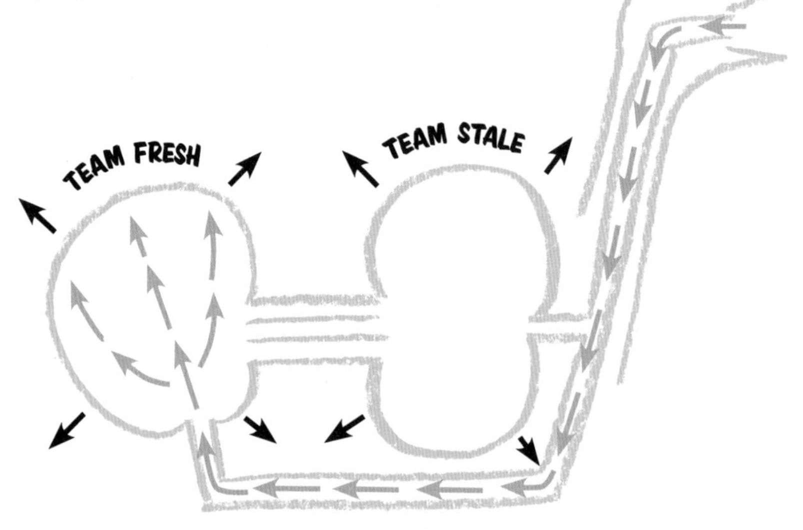

STEP THREE: PUFF #2

As Eno breathes in again, Team Fresh fills with fresh air from the outside world, just like in Step One. At the same time, Team Stale's balloons fill up, not with fresh air but with stale air pulled from the lungs.

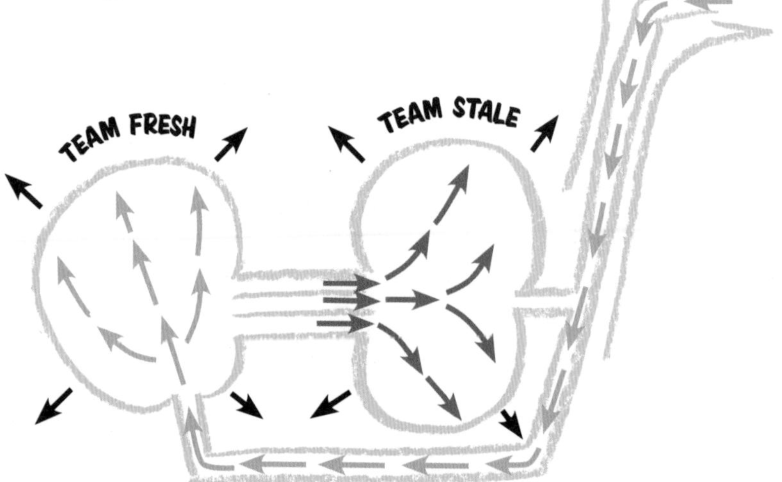

STEP TWO: SQUASH #1

As Eno breathes out, Team Fresh doesn't push air back out into the world, as your lungs would. It pushes air into the bird's parabronchi. (Whoops! I said I'd call them tubes.) These tubes are Eno's lungs. And, just like in your lungs, this is where blood picks up oxygen and drops off carbon dioxide. So, just to be clear, although Eno is breathing *out*, fresh air is flowing *into* his lungs.

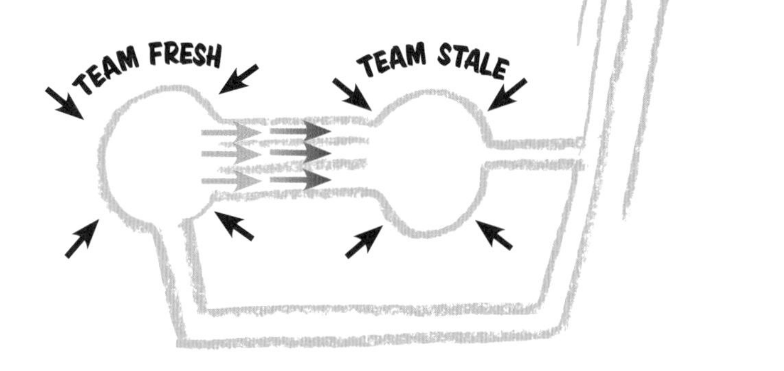

STEP FOUR: SQUASH #2

As Eno breathes out again, Team Stale squashes all the stale air out of Eno. (Also, Team Fresh squashes fresh air into Eno's lungs again.)

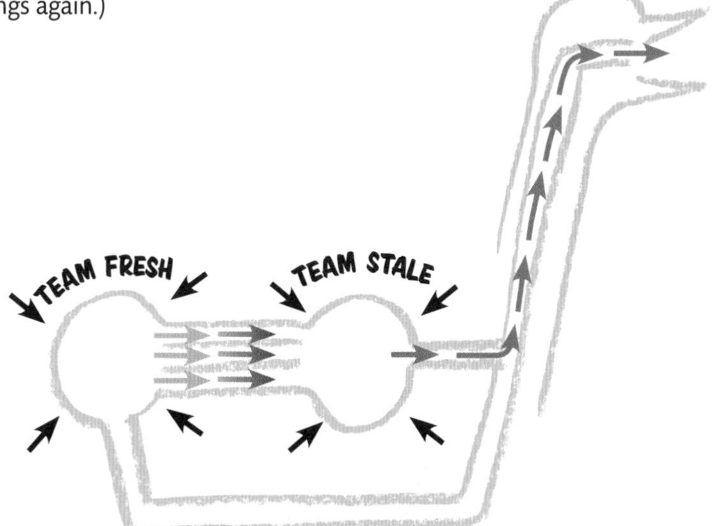

So, instead of our simple in and out, birds breathe in four steps: air goes into Team Fresh, across the lungs, into Team Stale, and out. And that means birds don't breathe out the same air they just breathed in—it takes two breaths in and two breaths out to pass air through the whole lung system. It also means that fresh air is always flowing through Eno's lungs whether he is breathing in or out!

AMAZING! STUPENDOUS! WHIZ-BANG-WOW!

But what's even more amazing is what birds can do with those lungs.

SUPERPOWER #8

EPIC ENDURANCE

BIRDS ARE HIGH-ENERGY ROCKETS that need as much as twenty times more oxygen than humans. Good thing bird lungs are ten times better at sucking every last bit of oxygen from the air. There's no way human lungs could keep up with a bird's supercharged lifestyle! **THE IMPOSSIBLE EVER-FLOW LUNG** is what allows birds to fly for days or soar at super-high altitudes where there's barely any oxygen at all. And those amazing lungs give ostriches **EPIC ENDURANCE!**

Cheetahs are superfast sprinters, but they get winded after a minute of high-speed chasing because their mammal lungs just can't get enough oxygen to power their muscles for long. Eno may not be able to sprint as fast as a cheetah, but with rocket lungs fueling his muscles he can run at incredible speeds for an incredibly long time.

In fact, ostriches are the world's fastest distance runners! Eno could run a marathon in forty minutes flat. That sounds impressive, but let me explain how truly amazingly fast that is. If the world's fastest 100-meter runner sprinted an entire marathon at a 100-meter pace, Eno would still be way faster . . . by about half an hour!

And speaking of **EPIC ENDURANCE,** you've got some **EPIC GRIT** yourself! That was a lot of lung info. But you made it through with flying colors. Yay for you! **AND HERE IS YOUR PRIZE:** I guarantee you now know something your parents don't, unless your parents are biology teachers or veterinarians. In fact, I guarantee no adult knows a darn thing about the avian lung, except those biology teachers and vets.

Why not put your parents to the test? Over dinner tonight say something like, "Hey, mom and dad, did you know that ostriches don't breathe out the same air they just breathed in?"

I bet they'll just blink at you because what you've just said is impossible, for a human. Hopefully they don't say something boring like, "That's nice, dear." I hate it when people say such things when I've just told them the most amazing animal fact.

And here are three more amazing but true ostrich facts to discuss at dinner tonight.

AMAZING BUT TRUE

1. There are actually two species of ostrich, and one of them is blue.

It's true! For centuries, scientists thought there was only one species of ostrich with several slightly different-looking subspecies. But now they believe the Somali ostrich, also known as the blue-necked ostrich, is a separate species. And yes, the male Somali ostrich's neck and legs turn blue in mating season. It lives in eastern Africa, in Ethiopia and Somalia.

OSTRICH FACTS

2. Johnny Cash, the legendary country singer, was almost killed by a *TOE CLAW OF DEATH!* If it hadn't been for his belt buckle, he surely would have died.

It's true! Johnny Cash built himself a zoo in Tennessee. One day while he was walking, an angry ostrich attacked. A kick from the ostrich broke two ribs and tore Cash's stomach open to his belt buckle. When Cash fell, he broke three more ribs on a rock. He grabbed a stick and hit the ostrich in the leg, so it ran away instead of stomping on him. Nasty business!

3. You might know ostrich feathers make good dusters. But did you know some car factories use thousands of ostrich feathers to dust cars before painting them?

It's true! The Ford factory in Valencia, Spain, is a high-tech robotic wonderland, but nothing is as good as ostrich feathers for getting rid of dust. Before getting a paint job, each car is gently dusted with enormous feather-covered robotic rollers to make sure not a speck of dust remains. Unlike other dusters, which just move dust around, ostrich feathers become charged with static electricity (which is like a magnet for dust) when they are rubbed together. (The feathers all come from ostrich farms.)

SAVANNA SURVIVAL KIT

YOU KNOW A LOT ABOUT OSTRICHES NOW. You know they have **COLOSSAL ORBS OF TELESCOPIC VISION** for spotting hungry lions. They have **THIGHS OF THUNDER** and **SUPER-FANTASTIC ELASTIC STRIDERS** for running at supersonic speeds. And don't forget those **TOE CLAWS OF DEATH.** In other words, you know ostriches are built fast and fierce to survive the savanna.

So how does an egg survive?

Most birds lay their eggs in trees for safety. Ostriches lay their eggs on the ground. Most birds that nest on the ground lay brown or speckled eggs for camouflage. Ostrich eggs are pure white and enormous.

So how does an enormous glowing white egg survive in the hot savanna dirt?

Well . . . it doesn't. (I bet you didn't see that coming. Or maybe you did.) I'm sorry to say so, but it's true. More than 80 percent of ostrich nests are raided by baboons, hyenas, vultures, lions, and other hungry beasts. And less than 15 percent of chicks survive to their first birthday. Yikes! I told you life is tough here. But it makes the eggs that do survive all the more impressive. Introducing **THE EGG OF WONDER!**

SUPERPOWER #9

THE EGG OF WONDER

YOU PROBABLY KNOW THAT OSTRICH EGGS are the biggest eggs around, but did you know their shell is so thick it can hold the weight of a fully grown man? Or that you would need two dozen chicken eggs to make a bigger omelet?

*ACTUAL SIZE

But none of that is important to ostriches. What is important is that their eggs survive. At least, it becomes important. Ostriches have a weird way of going about parenthood at the beginning.

Most birds will lay a few eggs in a nest and immediately sit on them for warmth and protection. Ostriches think they have a better system. First, the male ostrich will dig a shallow pit in the dirt. Then, the ladies (in this case Uma, Ada, Astrid, and Fadhila) will each lay one egg every few days in the nest-pit. Or anywhere else: ostriches might lay an egg in the grass, on a pebble, in a hole, over here, over there, or in some other ostrich's nest. Perhaps Uma saw a lion just as she was about to lay her egg and ran away. Or maybe a baboon had raided the pit—ostriches won't lay in nests that have been raided. In any case, ostriches do try to lay their eggs in their pit, and the ladies may lay as many as forty eggs over several weeks.

And guess what they do after laying each egg?
You'll never guess.

They just walk away!

That's right—ostriches lay an enormous white egg in the savanna dirt, then just leave it alone for weeks! No wonder 80 percent of eggs don't survive! Now, the superstrong shell is too hard for many animals to break open. And while glowing white eggs are easy for hyenas to spot, the white actually saves the eggs from cooking. A brown egg would soak up so much heat from the sun, it would hard-boil before Uma returned.

I don't know why Uma finally decides to return and protect her eggs, but when she does, she's ferocious! First, she kicks out any extras—she can only really sit on twenty or so eggs. (Scientists are almost 100 percent sure that Uma can't tell which are her eggs.) Then she sits and sits and sits in the broiling savanna sun for six long weeks. Actually, Eno and Uma take turns sitting on the eggs. Uma during the day and Eno at night, when his black feathers help camouflage him.

The old myth that ostriches hide their head in the ground when danger is near isn't true! But if Uma can't escape danger, particularly when she's guarding eggs, she might lay her neck flat on the ground. Since her neck and head are the same color as the sand, she is perfectly camouflaged and it might look like her head is buried. Also, ostriches peck the ground some 30,000 times a day. From a distance, it might look like their heads are hidden in a hole.

When the baby chicks hatch, they'll be as big as full-grown chickens, with beige and speckled feathers. The speckles are spectacular camouflage, but the chicks still look like supercute chicken lollipops. Guess who likes chicken lollipops? Lions, hyenas, mongooses, cheetahs, big snakes, jackals, African wild dogs . . . I could go on. Which means all of Eno and Uma's **SUPERFIERCE SURVIVAL SKILLS** go into high gear.

Eno will stand tall and watch with his **COLOSSAL ORBS.** Uma will chase down a hyena and stomp on it. She'll kick that lion right out of town with a **TOE CLAW OF DEATH.**

Ostrich parents show no mercy when they have chicken lollipops to protect.

Luckily, the chicks don't stay little for long. By the time the chicks hatch, the rainy season will have begun again, and the grasses will be tall and luscious. Chicks grow nearly a foot (25 centimeters) each month, and by their first birthday, they'll weigh as much as a baby rhinoceros!

But the babies haven't hatched yet. It's still late September. Uma is still sitting on her eggs. And it still hasn't rained for weeks and weeks. Time to talk about that other killer in the Serengeti: the hot, hot baking sun and not a drop of water.

HOT! HOT! HOT!

THE TEMPERATURE DOESN'T CHANGE MUCH between the wet season and the dry season, but when it hasn't rained for months and the watering hole has shrunk to a dirty puddle, you can be sure the sun seems a whole lot bigger, hotter, and meaner.

Now, I've told you that Eno can cool off a bit by lifting his **DO-IT-ALL DINO FLAPS** off his naked thighs. But that's sort of like taking off your sweater in summer—you're definitely cooler, but the hot sun is still baking down on you. If the temperature creeps hotter, you might get heatstroke. That's when your body heats up to 104°F (40°C). And that can kill you!

To cool off, you can find shade, pour yourself an icy drink, or turn on the fan. If you're still hot, you'll sweat, which is your body's way of cooling off. But Eno can't do any of those things. There isn't any shade. The watering hole has dried up. There's nowhere to plug in a fan. And birds don't sweat.

And it gets worse! During the dry season, the temperature can reach 122°F (50°C). Boiling hot! Red-deadly hot! But does Eno get heatstroke? Does he sizzle up like a deep-fried chicken wing? Of course not! He's got a **HYDRO-HOARDING HEAT SHIELD** to protect him.

POOP ALERT! Dirty business ahead. But it won't be too gross. Definitely not as gross as that time a seagull pooped on your sister's head.

SUPERPOWER
#10
HYDRO-
HOARDING
HEAT SHIELD

FROM TOP TO BOTTOM, Eno is perfectly designed to survive horrible heat and dryness. In fact, Eno's **HYDRO-HOARDING HEAT SHIELD** is built right into his name. Scientists call ostriches *Struthio camelus*. *Struthio* is Latin for ostrich, and *camelus* is Latin for camel. And like camels, Eno can survive weeks without drinking—even during the driest, hottest weather. And here's how he does it.

When it gets hot, Eno pants, just like your dog. Except Eno is an **OLYMPIC-LEVEL PANTER.** Panting works a bit like sweating. As you sweat, the water on your skin evaporates into the hot air, and that cools you down. Same thing with panting. As Eno pants faster and faster, the moisture inside his throat evaporates as hot air hits. Evaporation happens all along Eno's extra-long neck, which means superpowered cooling.

But there are two big problems with panting.

PROBLEM #1, HYPERVENTILATION: that's when you breathe in and out so fast, the air inside your body gets out of balance. If that happens, you get dizzy in the head, your muscles stop working, and you might pass out. But this isn't any sort of problem for Eno. He has a special exhaust pipe that allows him to breathe in and out without letting air into his lungs at all.

And then there's **PROBLEM #2, DEHYRATION:** that's when your body doesn't have enough water. The more Eno pants, the more water evaporates, and the drier and drier he gets. Eno gets some moisture from roots and grasses, but that wouldn't be enough for you. So what's an ostrich to do?

ACTIVATE THE HYDRO-HOARDING HEAT SHIELD!

The real power of the shield is to make sure whatever water goes in the top doesn't come out the bottom, if you know what I mean.

Now, all birds (including ostriches) only have one opening down below. It's called a *cloaca,* and only one thing comes out of it: wet squelch. Squelch isn't only gross, it's a big waste of water. Which is why ostriches are the only birds that pee and poop! Ostriches just keep getting weirder and more amazing, don't they?

Ostriches have superpowered kidneys that recycle most of the water from their pee back into their bodies. When ostriches get dehydrated, their pee becomes white and sticky. Super gross but super hydro-hoarding! And ostriches have long intestines that suck all the water from their poop before it leaves their bodies, so what comes out are hard, dry knobbles. **NOT A WATER-WASTING SQUELCH IN SIGHT!**

But enough about all that. Time to put Eno's superpowered parts together!

INTRODUCING
ENO:
SUPERSONIC
SURVIVOR!

THERE ISN'T A BIT OF ENO THAT ISN'T DESIGNED to survive the meanest, hottest, thirstiest, fastest enemies in the savanna. Extreme speed? Eno's got it. Epic endurance? Got that too. Telescopic neck? Colossal eyeballs? Killer kick? Water-saving poop knobbles? He's got it all!

But of all his superpowers, which do you think is the most powerful?

Can't decide? That's because it's a trick question. What makes ostriches truly superpowered is how all their superpowered parts work together. It's not just that Eno has **THIGHS OF THUNDER** or **SUPER-FANTASTIC ELASTIC STRIDERS.** It's how those mighty thighs and elastic legs work with his horrible feet, rudder-wings, rocket lungs, and even his wiggly neck and tiny head to turn an ordinary ostrich into a **SUPERSONIC SURVIVOR!**

If you had a shorter neck or a bigger head, you would still survive in this world. But ostriches are built differently—Eno and Uma are perfectly designed survival machines built to conquer the punishing, high-speed savanna, where life or death is always just a second or an inch away. But they're not scared. Ferociously fierce! Unfathomably fast! These battle birds can conquer anything or anyone. (Except maybe math.)

But if you think Eno is amazing, you should hear my story about **OLENKA THE EEL, MIGRATING MISTRESS OF MYSTERY!**

GLOSSARY

ALVEOLI (al-VEE-O-lie): Tiny balloon-like air sacs that make up your lungs.

AVIAN (ay-VEE-an): A fancy word for bird.

BINOCULAR (bi-KNOCK-cue-lar) ***VISION:*** What both eyes can see together. It's useful for judging distance.

BOLUS: A lump of food or medicine that is swallowed all at once.

CARBON DIOXIDE: One of the gassy building blocks of our world. It is what you breathe out and what plants breathe in.

CLOACA (klow-AY-ka): A single opening down below for both kinds of business. Most animals (except most mammals) have one, and what comes out is squelchy.

CORNEA (corn-KNEE-ah): A clear layer at the front of your eye that protects your eye and helps focus light entering through your pupil.

CROP: A throat pouch where food collects.

DEHYDRATION (DEE-hi-DRAY-shun): When your body doesn't have enough water.

FOVEA (FOH-vee-ah): A tiny pit in the retina that gives the clearest vision.

GIZZARD: Muscular stomach pouch for grinding up food. Ostrich gizzards are full of stones.

HYPERVENTILATION: When you breathe so fast, your body gets rid of too much carbon dioxide. It makes your blood too alkaline, your head dizzy, and your muscles weak.

KAPUT (KA-put): German for broken, useless, no good, all washed up, and down the drain.

LENS: A clear oval lump inside the front of the eye that changes shape to focus light on your retina.

OXYGEN: You can't see it, but it's all around you, and you need to breathe it to survive.

PARABRONCHI (PEAR-ah-BRONK-eye): Bird lungs.

PECTEN: A lumpy thing in bird (and some reptile) eyeballs that helps nourish the retina, reduce glare, and generally keep bird eyes sharp and healthy.

PERIPHERAL (per-RIF-er-al) ***VISION:*** What you can almost see at the sides while looking straight.

PHOTORECEPTORS: Light-gathering information units in your eyeballs. There are various kinds. Humans have rods, which collect data about light, dark, and movement, and cones, which collect color data. Birds also have double cones, which are doubly cool.

PUPIL: The black hole at the front of your eye that lets light into your eyeball.

RATITES (RAT-ites): A group of flightless birds including ostriches, emus, rheas, cassowaries, and kiwis.

RETINA: The screen at the back of your eyeball.

SAVANNA: The African savanna is a grassy belt across the middle-ish of Africa that gets just enough rain to stop it from becoming a desert.

TENDONS: Tendons attach muscles to bones so muscles can move your body parts.

THEROPODS (THAIR-o-pods): An ancient animal group that includes birds and a bunch of very fierce dinosaurs.

VITREOUS HUMOR (VIT-tree-us HUE-mur): The jelly in the middle of your eyeball.

UNGULATES: Animals with hooves, like gazelles, horses, cows, and rhinoceros.

FURTHER OSTRICH READING

If you have made it through this book, you're done with ostrich books written for kids. On to the big leagues for you! Here's a book with lots of great historical images you might enjoy:

Ostrich by Edgar Williams (Reaktion Books, 2013)

IF YOU WANT TO KNOW ABOUT SERENGETI CONSERVATION:

Frankfurt Zoological Society—Serengeti Conservation Project
fzs.org/en/projects/serengeti-conservation

AND YOU CAN NEVER WATCH TOO MANY OSTRICH VIDEOS:

ARKIVE: Lots of facts, photos, and videos:
www.arkive.org/ostrich/struthio-camelus

And if you've ever wanted to see an ostrich in a bicycle race:
www.youtube.com/watch?v=unqOeybMVeY